VOICES FROM A FUTURE PASSED

HOW THE BRITISH BROADCASTING CORPORATION, ACORN COMPUTERS AND ARM CHANGED THE WORLD

ROBERT NAPIER

doit:once
PO Box 97, Nicholson VIC 3882 Australia
www.doitonce.net.au

Title: Voices from a Future Passed

First published in 2026.

ISBN: 978-1-7644608-0-4 (print)

Cataloguing-in-Publication data is available from the National Library of Australia.

Creator:	Napier, Robert G, author and designer
Cover photo:	Edward Pedersen

Acknowledgement: Trademarks, trading names, brand names and product names referred to in this book are the property of their respective owners and copyright holders.

Dedication

If there is a lesson to draw from
the stories told by the contributors
to this book, perhaps it is this:

The past is life's classroom: Learn all you can from it.
Savour the present: Make every moment count.
Whatever comes, face the future with confidence.

To Hamish & Leo

We are all made from stardust,
with just a pinch of pixie dust!

Foreword

I have a vivid memory of saying to an able young man, David Allen, that there may be something in this microelectronics thing, and worth looking into. In retrospect, it seems ridiculous; at the time it was prophetic.

That casual request led to the *Computer Programme*, broadcast on BBC British television in 1982. It was a world-first initiative that addressed the potential impact of the emerging field of digital electronics; but that was just the beginning. Other award-winning BBC television programmes followed well into the next decade, influencing generations of viewers.

With vision, skill and dedication, the *BBC Computer Literacy Project* team, with support from the Thatcher Government, worked with a small but impressive company in Cambridge – Acorn Computers. British industry and the BBC's viewing audience were the immediate beneficiaries.

While I don't pretend to understand digital technology, who can fail to appreciate the remarkable changes it's made in every aspect of our lives?

Rob Napier's well-researched history, *Voices from a Future Passed*, brings the story to life through the recollections of people at Acorn Computers, the BBC and the University of Cambridge who were directly responsible for bringing a virtual revolution to Britain and the rest of the world.

The BBC is one of the world's most respected broadcasters. This doesn't happen by accident. It is the result of courageous people in every part of the organisation working tirelessly to create quality programmes of which the *BBC Computer Literacy Project* is just one example.

Sheila Innes
Former Controller of Educational Broadcasting
British Broadcasting Corporation

Preface

Voices from a Future Passed tells a story about rapid change – economic, social, political, technological. It began in the USA and the UK, as new uses for transistors and digital computers appeared daily, with its effects now felt across the globe.

This story is presented by the people who were there. Early contributors to the information age describe the highs and lows, the wins and losses that have affected all of our lives – and will continue to impact the lives of future generations. Working with Acorn Computers and the British Broadcasting Corporation (BBC) in the early 1980s was an adventure. These bright independent people were loyal to their colleagues and committed to shared goals. These attitudes were never seen as being virtuous – they were fun.

"Acorn Computers is what happens when there is a concentration of really, clever people who happen to be in the right place at the right time and have the courage to take advantage of opportunities.

"They came together almost by accident. They weren't there because of their business ability, so lots of things did go wrong. Everything relied on individual motivation and a deep personal sense of responsibility. No one tried to control anyone. It was a very open environment. You were expected to see what needed to be done and get it done. Most of the time that worked spectacularly. When it failed, it was equally spectacular."

David Bell, Acorn's Technical Director 1990–1995

Despite the technological advances since the 1980s, retro computing enthusiasts are keeping alive an interest in what Acorn Computers achieved; but published information, even from credible sources, can be factually wrong. The internet provides fertile ground for misinformation peddled as facts. To avoid retelling those myths, most details are supported with documentary evidence, or if not available, by oral history from no fewer than two reliable sources – people who were present at the time. Even these precautions don't guarantee 100% accuracy as sometimes there are disputed recollections.

A lot has happened since the 1980s. Memories have blurred, but solid evidence was found for what is presented here. A few recollect events from different perspectives. You are left to decide which is most likely: Is it one, or the other, or perhaps both? After intensive research, this biography may be as close to certainty about the history of the *BBC Computer Literacy Project* and Acorn Computers as is possible to achieve.

The source material was gathered from BBC correspondence, face-to-face interviews, chat sessions, online resources, notebooks, emails, and even Telex printouts – as well as assistance from the National Museum of Computing, the Centre for Computing History in Cambridge, and the Computer History Museum in Mountain View, California. These, and other museums around the world, are preserving this significant phase of human development.

Much of this book is written in the first person, reflecting on the experiences of people who were involved. In each case, the final choice of words is mine, but the ideas and intent belong to the named sources. I set out to speak to as many people as possible. Sadly, a few on my wish list have passed away and a few couldn't be reached. I'm grateful to those who have allowed me to tell this remarkable story – warts and all.

This work is divided into seven sections:

- BBC producer, David Allen, and BBC engineer, Richard Russell, discuss the *Computer Literacy Project* and the appointment of Acorn Computers to build the BBC Microcomputer.
- Acorn directors, Christopher Curry and Sir Andy Hopper, and one of its Managing Directors (after the Olivetti takeover), Sam Wauchope, discuss Acorn Computers and the Olivetti Research Laboratory.
- Acorn's technical dominance, as told by three people who led development of the BBC Micro, the ARM processor, the Archimedes, and other Acorn systems: Chris Turner, Steve Furber and Sophie Wilson.
- Sixteen engineers, scientists and mathematicians who worked in Cambridge and in Palo Alto offer unique perspectives. Combined, they build a clearer image of the company throughout its existence.
- The Australasian experience – how it affected Acorn development.
- In the best traditions of the Brother's Grimm, there is a fairy-tale ending for Acorn shareholders as the company was wound up.
- The Further Reading chapters include explanatory notes and other useful background information.

Contents

Foreword
Preface
Contents

00000000	Prologue	1
00000001	Barking Mad	5
00000010	David Allen	11
00000011	The Langham Decision	23
00000100	Richard Russell	29
00000101	Sinclair v BBC v Acorn	33
00000110	Christopher Curry – before Acorn	47
00000111	Christopher Curry – Acorn harvest	57
00001000	Andy Hopper	71
00001001	Sam Wauchope	79
00001010	Chris Turner	83
00001011	Steve Furber	99
00001100	Sophie Wilson	111
00001101	Andrew Gordon	125
00001110	Arthur Norman	135
00001111	Brian Cockburn	143
00010000	Carl Dellar	151
00010001	Colin Priestley	159
00010010	David Bell	165
00010011	Jes Wills	171
00010100	Jim Mitchell	177
00010101	Joe Dunn	181
00010110	John Cox	191
00010111	Jon Thackray	195

00011000	Kim Spence-Jones	205
00011001	Laurence Hardwick	209
00011010	Paul Fellows	217
00011011	Ramanuj Banerjee	227
00011100	Acorn Downunder	233
00011101	A Fairytale Ending	241
	Further Reading	
00011110	Special Mention	249
00011111	The Radcliffe Report	253
00100000	Before the Revolution	259
00100001	BBC BASIC – A Primer	269
00100010	Cambridge v Manchester	275
00100011	Acorn Machines	279
	Acknowledgements	290

IT HAS BEEN SAID: There are 10 types of people in this world – those who understand binary numbers and those who don't. In keeping with the theme of this book, chapters are numbered in binary notation.

00000000

Prologue

Our story begins at 4A Market Hill in Cambridge, where graduates, students and researchers – obsessed with digital electronics – designed and built computer hardware and wrote software that could run on these tiny but capable machines. This team of engineers, mathematicians and computer scientists launched one of the UK's most successful startups.

On the back of a project to re-design a mechanical gaming machine in the late 1970s, Christopher Curry and Hermann Hauser contributed just £50 apiece to launch a consulting business – Cambridge Processor Unit. They were in a hurry to make their fortunes and decided that charging by the hour would never achieve that.

Manufacturing seemed to offer more opportunities, so they registered a second business, Acorn Computers. That decision coincided with events that increased the value of their investment by about one million times in just five years. But fate has ways of playing tricks on all of us: Two years later, there was a fire sale. The founders sold the business for a fraction of what it had been worth. The roller coaster ride continued, with the final upswing reaching dizzying heights in 1999.

Acorn sprang to life in the Thatcher years. The British Broadcasting Corporation (BBC) generated a flood of interest in computing when the *BBC Computer Literacy Project* broadcast its first programme on UK television in 1982 – and later in the USA and British Commonwealth countries.

A year earlier, Acorn had been awarded a contract to produce the BBC Microcomputer. The BBC's evaluation team was keen to find a machine that fitted their vision and people who could deliver what they promised – and preferably British. They were all impressed with what they saw.

What the BBC's preferred vendor hadn't achieved in a year Acorn produced in four days. While it wasn't a complete design, they saw enough to know that these people could build what they needed.

There would soon be a BBC Micro in a million homes and classrooms – fifty times more than forecast. Acorn's *secret sauce* was the trust the founders placed in their people. Hermann Hauser and Christopher Curry created an environment that encouraged everyone to strive to reach their potential. Many built successful careers: some received national and international honours and awards; and a few became wealthy by commercialising their inventions.

Acorn became Britain's fastest-growing business. Starting with a handful of people, they soon employed hundreds – the catalyst for an industry that employed thousands in retail stores, service businesses and the media. It grew so fast that it reached a point where the company's human and financial resources couldn't control what was happening around it.

Their subcontractors failed to deliver on time. Product reliability called for after-sales support that couldn't match customer expectations. Without decisive change a downward spiral was inevitable. The BBC suggested ways that Acorn's directors could manage the growing pains – while protecting its own reputation. In time, Acorn overcame its production and support issues and began to ride a wave of success.

In April 1983, the founders floated the company. This injected extra capital. New managers with different experience joined Acorn – but it wasn't enough, so the company began to lose its way.

When International Business Machines (IBM) entered the personal computer market in 1981, it presented no immediate threat to Acorn, as the IBM PC's price point was more than most British consumers could afford – but that would change. As cheap copies from Asia (PC clones) flooded the market, customer sentiment changed too. The *BBC Computer Literacy Project* had enabled the public to become more tech savvy; so just being 'British Made' was no longer enough to close a sale. Acorn wasn't prepared for this new reality.

Despite its early successes, Acorn found itself pitched against IBM PCs and clones, as well as Apple, Commodore, Sega, Atari, Tandy and several UK manufacturers like Sinclair, Research Machines, Apricot and Amstrad. Worldwide, buyers were spoiled for choice.

Acorn seemed oblivious to the gathering storm as it made the ambitious decision to build its own microprocessor. Reflecting its strong research and development ethos, 1983 was the year a small team within Acorn, led by Steve Furber and Sophie Wilson, started a project that would create value far in excess of the rest of the business. It found new ways to think about the design of computer architectures – Acorn's Reduced Instruction Set Computer (RISC) microprocessor chipset was the result.

Project A – the Acorn RISC Machine (ARM) – was conducted in secret.

In the best traditions of a secret intelligence operation, those involved maintained complete silence until they were ready to launch.

Acorn tried to face external challenges head-on, but its resources were overstretched. By the end of 1984, the company had sunk millions into its premature attempt to enter US and Canadian markets; and the Electron, BBC Model B+ and the Cambridge Workstation had failed to achieve the popularity of the BBC Micro. The BBC Master 128 was launched in February 1985 as a fightback measure.

Acorn was haemorrhaging cash. In February 1985, as it faced serious liquidity problems, Olivetti offered them a bail-out by injecting about £12 million in cash. Acorn's directors accepted the offer as an opportunity to reduce debt and to empty its overstocked warehouses by tapping into Olivetti's worldwide sales network. If the directors understood the full extent of their problems, they didn't declare it.

As part of the agreement, Olivetti installed its own people in Cambridge to look more closely at Acorn's operations. By the time they understood the true financial state, they were committed. Clearly, more funding was needed. After some not-so-subtle persuasion from Government, Olivetti agreed to provide that funding. They took control of the company, installing some of their own Directors and the Chairman of the Board. Major creditors accepted large write-downs. The stock value of Acorn Group plc plummeted; but then, a miracle…

On 26 April 1985, the first ARM chip samples arrived and worked first time with barely an issue; and performed in ways never anticipated. Steve Furber found that the ARM1 consumed negligible power, ideal for portable computing. ARM would become the world's biggest selling processor design. More than 280 billion ARM ICs were sold worldwide over the next 40 years – more than 30 chips for every man, woman and child on the planet – and still with no sign of sales dropping. No one could have imagined the impact it would have on the world. Computers would never be the same.

Acorn wasn't the first to produce a RISC processor and wouldn't be the last. They were first to market with an ultra-low-power processor for mobile phones and embedded applications. Much of the electronic equipment we take for granted today is controlled by ARM technology.

Apple Computer wanted to use the chip in an early notepad device called the Apple Newton but was unwilling to use a component made by a competitor. Acorn proposed setting up a joint venture with Apple and VLSI Technologies (the chip manufacturer). This spinoff business, Arm Holdings, has been the most successful startup in Britain's history.

With its innovative licensing model – the brainchild of Robin Saxby – RISC chips based on the ARM processor became a runaway success. Offspring of the first ARM chip are everywhere. If you are not under the influence of at least one of their devices today, you should check your pulse!

Arm Holdings totally eclipsed Acorn's earlier successes, as interest in ARM technology soared around the globe. The stories presented here explain how it started; how its participants were drawn into the melee; and how they dealt with the challenges of the time.

00000001

Barking mad

To reach 4A Market Hill, Cambridge in 1981, you passed through an unimpressive entrance beside the Eastern Electricity showroom at the Market Square. The end of the passageway was so poorly lit, even the graffiti was difficult to read. A door into the building passed by the mail room where Dave Bard fought a never-ending struggle to cope with Royal Mail deliveries, which mostly contained computers returned for repair. A second, heavier door led to an electricity substation that shared the ground floor. The low hum of its large transformers was barely noticeable, except in winter when demand peaked; then the noise sounded like a drumroll.

Stairs led to the first of two floors, both occupied by Acorn Computers. On the first, Christopher Curry and Hermann Hauser had shared an office on the right, but with a continuous need for more space, they recently moved with the admin staff to 27 Bridge Street. Glyn Phillips and a few others had moved into their old office to set up a repair workshop.

On the left was a room filled with boxes, packing material, circuit boards waiting for components, and Colin Priestley's office. Despite the freezing weather, Colin and his staff assembled, tested, repaired, packaged and shipped from this building. The top floor housed the programmers and engineers designing new products. The bitter cold made it hard work for everyone. What they'd have given for central heating!

Though usually cheerful, Colin often seemed to carry the burden of the daily grind, and was quick to express his opinion, whether anyone wanted it or not. He was often heard ranting about parts arriving late – a regular occurrence. Component delays were the bane of his life. Soon, Colin's team would move to French's Mill easing the pressure on space – though not for long. In total, 12,000 Acorn Atoms were sold.

It was the first Monday in February. The weather was cold and miserable. As Colin planned his day, he could hear the faint buzz of excitement in Chris Turner's office on the top floor.

"They've started early," he noted. That was normal for Chris. Even in the middle of winter, it would take a national disaster to stop him arriving by eight o'clock. Hermann and Christopher, Sophie Wilson and even Steve Furber were crowded into his office, deep in conversation. Steve was Rolls Royce Research Fellow at Emmanuel College. The timing of his visits was unpredictable, but seldom this early, particularly on a Monday morning. Yes, something was up.

Hermann started the meeting, predicting that if successful, this could be the biggest thing to ever happen to them. He admitted it was a lot to ask, but if successful, the rewards would be huge. He finished by promising to give them all the support they needed.

Hermann and Christopher sounded bright and cheerful. Chris Turner was restrained. Sophie's and Steve's comments were muted. Hermann had phoned them both the night before, saying he wanted to build a prototype of their new design in four days. Their responses were similar: "Anyone who thinks we can do this in four days is barking mad!"

Hermann had been conservative with the truth, making each believe that the other had already agreed to the plan. Sophie and Steve knew they'd been duped but it didn't matter; they were committed. Later, they both admitted that they liked the idea of attempting this, even though it was impossible and sure to fail.

They weren't afraid of hard work. Four days of intense effort could win a contract to build thousands of computers. This could be a once in a lifetime chance. It was worth the effort and there was no real downside.

Christopher admitted that this plan was his idea. The Atom wasn't good enough to convince the BBC to let them build their computer. They needed to show them a new design – the Proton. They'd been talking about this for a while but now it was time to stop talking and start building. To have it ready by Friday was a long shot; so 'barking mad' really did seem appropriate, but worth a try.

Chris Turner offered the use of his office for the week. Steve and Sophie expected to take two 'big' days to complete the design. They could get the job done, but with no time to spare for niceties. Chris promised to source any components they needed. He'd ask Hitachi for some fast single-supply memory chips, but they were new and would be difficult to get.

Someone suggested Ram Banerjee should do the wire wrapping. They all agreed that no one was faster or more reliable with a wire wrap gun – the best man for the job. Hermann promised to use his powers of persuasion to recruit Ram, *the fastest gun in the west.*

No more discussion was needed. They all knew what had to be done. Chris disappeared and returned with large sheets of drafting paper and taped one onto his drawing board. He would draft the schematic as Steve's design evolved. He slipped out again to check component stock levels in the store.

Video displays on most home computers were of low quality because the controller couldn't access the video display memory each time the central processor wrote to it, causing short flashes of light. The effect was annoying, but Steve and Sophie found a way for the processor and graphic display to share the video memory without interfering with each other. It would need high-speed memory chips; and Hitachi was the only chip maker that made them.

Chris called his sales rep who drove up to Cambridge and hand-delivered the only eight high-speed RAM chips in Hitachi's UK warehouse. Its support for this fledgling business was excellent. Although the design wasn't finalised, Chris had a clear idea of what would be needed, and he couldn't afford to wait. By the time he called a few more suppliers, he had arranged for all parts to be delivered by the end of the day.

Steve and Sophie piled their data books and working notes onto a small table, then set to work. For their first pass, they developed the design on a sketch pad – outlining what the machine would need. Sophie prepared a list of ICs by function, then assigned part numbers to each.

They revised the outline a few times and by mid-morning, they were ready to start drafting the schematic: microprocessor, data and address buffers, decoders, memory chips, video controller, input/output (I/O) ports, and crystal clock – the building blocks of a digital computer.

Before spending too much timc looking at the connections, they stood back and considered what they had. This wasn't Steve's or Sophie's first design. They knew what to look for and found the obvious errors. Some of the not-so-obvious ones would show themselves as they drew the interconnections. Some would appear during wire wrapping, and the rest would remain in hiding until the prototype was debugged.

Electric heaters failed to overcome the biting cold coming in through gaps in the window frame. Hermann came in to check on progress, carrying mugs of hot tea and some *Mr Kipling Apple Pies* – always a favourite. He had phoned Ram, who promised to come at six o'clock the next evening.

They were making good progress but had a long way to go. The rest of the morning was more of the same: thumbing through data books for component timing, current loading and pinouts. It was slow, painstaking work but it was starting to take shape. They stopped for lunch, though this was punctuated with discussion about how the video interface would work. It was based on Steve's homebrew computer, but it needed a few changes. He thought about how to approach it over Christmas but hadn't reached a final decision. It was time to sort out the details, so he drew timing diagrams to test his theories. Sophie and Chris Turner checked his work and agreed with his approach.

By the end of the first day, the buses had been laid out on the drawing, connecting the CPU via the buffers and address decoders to the memory and I/O chips. Most of the other chips needed for the design had been roughed in so a final parts list was drawn up. Keyboard entry would be via a Eurocard System 3 using connecting cables. It was the only way the prototype could be built and tested in time. Chris checked the list and confirmed that he had all the components they'd need.

They worked until late, stopping just long enough to pop into the *Copper Kettle* for a quick meal. Finally, as they stepped out into the evening air after finishing work, they found it hard to breathe and ached from the biting cold. But so far, so good.

Tuesday morning, Hermann told the team that Allen Boothroyd would come in later in the day to discuss developing a mock-up of the case. By midday, pressure was starting to build. The schematic showed that the design needed about 120 chips. Checking the drawing would need every minute of the six hours remaining.

Steve and Sophie decided to skip lunch, but Hermann encouraged them to take a break. He ordered pizza from the *Eros* Greek restaurant in Petty Cury; and moved Allen's visit forward, so they could eat and talk at the same time. Allen learnt what he needed to know about the case, then left with Hermann and Chris Turner to tidy up the details. Sophie and Steve had watched the seconds tick by, though they'd stopped for only 20 minutes. It was time they couldn't recover, but the break helped, so they returned to the drawing board feeling refreshed and ready to press on.

The slow, painstaking work of checking each pin continued. The afternoon wore on, then darkness fell and the chill in the air reminded them that Ram would arrive soon. He was right on time. Without saying a word, Sophie slipped out of the building as the nightshift started.

Before wire wrapping could begin, Steve sketched the layout showing where to position each chip on the Veroboard.

Chris Turner had laid out bins of IC sockets, large sheets of Veroboard, a few hand tools and the battery-powered wire wrap gun. Ten reels were mounted on a bar above the workspace to dispense the coloured wire. Ram would connect each end of about 1,200 lengths of wire to socket pins – some 2,400 terminations.

Organising the support crew was like arranging deck watches at sea. The volunteers helped for a few hours at a time, reading pin numbers and determining which coloured wire should be used. This continued in shifts for 35 hours – only taking time out for a few short breaks. The pressure on Ram was brutal, but he stood up to the challenge with remarkable tenacity.

Wednesday was a blur. For each connection, they could take no more than 90 seconds, on average, to choose the start and end pin numbers, calculate the colour of the wire, load it into the gun, find the start pin, make the first connection, track the wire across the layout to the end pin, then cut and terminate it. Ram repeated these steps about one thousand two hundred times, until he had completed the prototype.

By five o'clock on Thursday morning, the wire wrapping was done. It was time to start debugging. Steve, Sophie, Hermann and a few others gathered twenty minutes before the last connection was made. Without fanfare or discussion, they were ready to start.

The cables from the in-circuit emulator were attached. The power supply was connected, and an oscilloscope was positioned to help with debugging. Steve fitted a few integrated circuits into their sockets, and someone said in mock humour, "Let the games begin." (They weren't chosen for their stimting repartee.)

Bleary eyed and painfully cold, they moved carefully across the prototype, populating each IC socket and checking for errors. Ram's work was almost perfect. A few wiring faults were found but most errors were due to problems with the circuit design. These were difficult to find, but one by one, they were eliminated. The shifts continued to change throughout the night and on into the early hours of Thursday morning.

Calling out the connections, the checker read out: "IC11 pin 14 to IC12 pin 14 – grey." Steve traced the pins, inspected the connections and the colour of the wire. "Correct," he responded. It continued this way until all connections were checked and found to be wired in, as designed.

As the debugging wore on throughout the day, fatigue set in. There was a feeling of desperation. Having come so far, no one wanted to leave. Night fell. More mugs of hot tea and this time, *Fitzbillies Buns* washed down the after-effects of a hot curry.

It was 2:00 am on Friday, the 6th of February. The BBC team was coming at ten o'clock that morning. Sophie said, "If we're going to get it working, it will need a program to run on it, and I'm the one who's going to write it. So, if you expect me to write something, I need to get some sleep." She left.

Some time around three o'clock in the morning, the entire circuit had been checked, yet still the ICE indicated the circuit was faulty.

Hermann, who was possibly the only person in the room able to think clearly said, "Well if there appears to be no fault, maybe there's a problem with the ICE. Why don't you plug in the microprocessor and see what happens?"

To everyone's annoyance, he was right. The board was working.

Sophie returned six hours later to find them all fast asleep on the floor. The oscilloscope display indicated the board was working, but it needed a demonstration program. She had two hours to modify the operating system to run on this new hardware, bootstrap BASIC, and get it running.

The BBC team arrived at 27 Bridge Street just before 10 o'clock that morning. Hermann and Christopher explained what the team had been doing all week. Meanwhile, Sophie had configured the communication functions. She programmed the new video controller chip using Atom BASIC statements to poke initial values directly into its registers.

When Hermann and Christopher brought their visitors into Chris's office, there it was: a high-resolution colour screen driven by the video controller on the Proton board. At first, it could only plot random lines across the screen; but it was working.

00000010

David Allen

Every revolution needs a trigger: A band of colonists, disguised as native Americans, tipped a load of tea into Boston Harbour, which led to the American Revolution. Some years later, crowds of angry peasants stormed the Bastille, starting the French Revolution.

The British Broadcasting Corporation is seldom thought of as being revolutionary but its actions in the 1980s contributed to rapid change for the people of Britain. It started by listening to the experts, doing its research, and reading community needs and opinions – then it enabled the groundswell that followed.

Before the internet, mass communication in the UK was BBC radio and television, with some input from independent broadcasters. It would be difficult to overstate just how much influence they had on the people of Britain. One of the nation's most trusted brands, the BBC continues to entertain, inform, and educate.

By the late 1970s it seemed that a 'microelectronics revolution' was coming and Britain needed to be prepared; but James Callaghan's Labour government showed little interest or understanding of the opportunities that new technology presented to industry and employment – and the dangers of inaction.

'Thought leader' is a clumsy buzzword, but in describing the BBC's role, it seems a good choice. It started with the BBC2 *Horizon* programme, *Now the Chips are Down*, first broadcast in the closing months of the Callaghan government. This government-funded broadcaster, openly critical of its masters, galvanized debate about what should be done to fix the UK's lack of awareness and competitiveness, anticipating the impact that new technology would have on Britain's workforce. In its closing scene, the last words were: 'The silence is deafening.'

The Manpower Services Commission (MSC) was tasked with advising government on employment policy. It decided to fund an independent investigation. The BBC's credibility and its international connections made it the ideal choice to undertake the work.

Head of Continuing Education, Television, Sheila Innes, appointed two BBC television producers, David Allen and Robert Albury, to investigate and report on the likely impact of microelectronics. The BBC had connections within some of the world's best research institutions, universities and corporations. After tapping into that network, David and Robert were convinced that BBC Television should help its viewers prepare for a future that would be dominated by computers.

This had a life-changing influence on David's career. As producer of the *BBC Computer Literacy Project*, he championed the need for action. His contribution, as well as and the work of others at the BBC, Department of Industry, in the education sector and Acorn Computers, helped to lift the level of computer literacy in Britain. This is David's story…

"It was late 1978. I was sitting in the bar at BBC Villiers House in West London, pint in hand. My boss, Sheila Innes, walked over to me clutching a glass of wine. At the time, I was facing the prospect of making a series that I didn't want to do. Sheila was about to throw me a lifeline. After some small talk she said 'The Manpower Services Commission has been in touch. They've asked us to look into a thing called microelectronics. Could you take a look and see if there's anything in it?' I jumped at the chance. It changed my life.

"I received my science degree from Balliol College, Oxford but my only exposure to computers was to take a group of students to see an analogue machine in the 1960s. When I typed 2x2, it answered 3.999! Digital computing showed more promise.

"With an allowance of £10,000 for the assignment, we went to Holland where Philips was developing 12" Laserdiscs and compact discs; then to Germany where the silicon chip was called the job killer, though workers felt positive about the future. We observed that having workers on the boards of companies helped make German industry more successful. Next stop was Sweden where unions were concerned about job losses, but newspaper production was already digitised, and car factories were using robots. In the USA we visited IBM's research centre, AT&T's cellular phone development, and then we met experts at MIT. Finally, in Japan we saw amazing, automated factories turning out robots and other new products. One of them was a beautifully-made handheld computer – the PC1500, launched by Sharp on the day we were there. It had one kilobyte of memory, a one-line display, and a primitive BASIC; but it started me thinking about programming. It's now in the National Museum of Computing at Bletchley Park.

"What we saw were technologies still years away from going to market, but the message was clear: the UK must learn to control technology, or risk being controlled by it.

"Our report detailed the results of the study – its descriptions, predictions and warnings. It outlined what technologists, industrialists, unionists and educators were saying, and helped shape thinking in government, industry and in the training sector. It also influenced BBC thinking about a hands-on public awareness and education campaign. Robert and I had formed strong views and Sheila supported us: The BBC needed to raise public awareness of this new technology and to help people embrace it.

"After the *winter of discontent* in 1978, when Britain faced rolling strikes, job losses, business closures, and record cold weather, there was a general election early the following year. The Conservative Party, led by Margaret Thatcher, came to power with a 44-seat majority in the House of Commons. *The Iron Lady's* time in office is associated with upheaval in Britain's heavy industries and the Falklands War, but her government also enabled the IT revolution in the UK – one that is still being felt today.

"Supported by the Thatcher government, planning for the *BBC Computer Literacy Project* started in 1979, and was launched in March 1982. With the resources of the BBC, and with government and the education sector behind us, it became an ambitious multimedia initiative. British industries got on board. This led to a rapid uptake of new technology, with thousands of jobs created that needed new hi-tech skills.

"The *BBC Computer Literacy Project* team was led by executive producer John Radcliffe. I served as editor and series producer, supported by BBC Engineers David Kitson and Richard Russell, and some external consultants including computer education evangelist, John Coll.

"As the project grew in importance, the BBC conducted audience research, as well as sizing up and preparing the education community. Paul Kriwacek was working on the first series – *The Computer Programme.* I was responsible as project editor for the television programmes; and though I was series editor, Paul was very much his own man, so I had little involvement. He got on with it while I looked after the publications that would be required for the project.

"BBC Adult Education was the obvious place to host this series as it had produced similar programs for professional groups and people with special interests. These were usually supported by publications or training courses. Our ambitions were quite modest, but the technology was evolving at such a rapid rate, and others were continually challenging our thinking, so we had to be flexible.

"I formed a strong opinion that the way to help viewers master this technology was to show them how to write programs for a variety of applications. I believed this should be done systematically and for that, we needed a versatile and TV-studio-friendly computer. Some disagreed, but this view prevailed. The idea of a BBC machine licensed by BBC Enterprises was controversial, but our proposals for computer literacy television and radio programmes resonated with the Government, which gave its tacit support – or at least did not object."

Television programmes

"During the project's planning stage, I produced *The Silicon Factor* with Bernard Falk to raise awareness. Launched in 1980, this three-part series for BBC2 mid-evening won an award at the *New York Film Festival*. Next, Brian Redhead presented a 10-part series: *Managing the Micro*, which explored the impact computers and technology would have on all of us, and how small businesses could use digital technology.

"During 1982 and 1983, four more television series aired: *The Computer Programme, Making the Most of the Micro, Computers in Control* and *The Electronic Office*. It may be hard to believe, but the material covered was new to most viewers. The programmes combined demonstrations of computers being used in business, industry and research.

"*The Computer Literacy Project* received the prestigious *Judges Award* from the Royal Television Society. By 1983, each programme had been watched by about two million viewers in the UK. They were broadcast by the Australian Broadcasting Commission (ABC), the Broadcasting Corporation of New Zealand (BCNZ) and the Public Broadcasting Service (PBS) in the USA. The four series were heavily used in schools, with colleges of further education recording and re-showing them."

Other Initiatives

"The BBC drew on its experience with the *Adult Literacy Project*. Its TV series, *On the Move* (featuring Bob Hoskins) provided reading courses for two million Britons, and reached a broader audience through education officers who engaged with schools and adult education centres.

"The project needed new programmes and learning packages. I talked to the National Extension College (NEC) about developing a training course and to a few other publishers. The NEC developed 30-Hour Basic in partnership with the BBC for people wanting to learn computing. More than 150,000 completed the course, making it one of the most successful NEC distance learning courses ever produced.

"The BBC produced books, courses, and software for reinforcement. I edited *The Computer Book* to accompany *The Computer Programme*. It sold 80,000 copies. This was considered a best-seller for a book at that time."

The BBC Microcomputer

"We had begun by thinking about a television series that explored digital logic – the world of gates, bits and bytes. This was soon overtaken by the idea of viewers learning to program computers at home. If you show things on TV and you want people to take part, you must have something they can try out for themselves. Otherwise, it would be like running a cooking programme where the viewers have no access to a kitchen.

"Computer literacy is more than simply knowing how to program, but since user software was so scarce that's how it was done. Games and other software for home use were still a year or two away. That led to the question of which computer would we use with the TV programme.

"There were some cheap microprocessor evaluation boards, but they didn't go far enough for what we had in mind. Some of the newer and more affordable machines coming onto the market like the Sinclair ZX80 and the Acorn Atom were released in 1980, and there were some more expensive American machines like the TRS-80, Apple II and Commodore PET. When I tested these machines, I found the versions of BASIC used were very different – even versions from the same supplier. We wanted to avoid a *Tower of Babel* effect with non-standard programming languages.

"With modest ambitions, we talked to Newbury Laboratories on advice from the Department of Trade & Industry (DTI). Newbury was funded through the National Research Development Corporation, so it was considered a suitable partner for the BBC. They inherited a prototype machine, the *Newbrain*. After a few months it became clear to us that they lacked the expertise to improve it. As our confidence in Newbury ebbed away, we co-opted John Coll in mid-1980 to advise us on the specifications for the machine we needed for the television series.

"John was Chair of Micro Users in Secondary Education (MUSE) and was teaching at Oundle School at the time. He proposed a 'structured' BASIC which MUSE promoted. Among other functions, he was adamant that the machine should be able to control and monitor things.

"The BBC had developed CEEFAX, so it was interested in Telesoftware. Also, BBC studio engineers needed a machine that would give very clear images when broadcast as part of their television transmission. These were our first requirements when we started to develop our specification.

"With no common version of BASIC among the available machines, there was a need for something structured and practical was needed to teach good programming skills. Also, there were few opportunities to connect external devices for performing process control experiments. There was nothing like today's Raspberry Pi back then; even the mouse and the compact disc were still a few years away.

"Ray Kernow from the Science Policy Research Unit at the University of Sussex joined us as a consultant in the autumn of 1979. Ray and I invited representatives from the DTI and manufacturers to a meeting in Cavendish Square W1. We presented John Coll's ABC – *Adopted Basic for Computers*. We hoped to convince suppliers to develop a well-structured compatible BASIC for all machines – including the one to be used in our television series. We also hoped this would placate manufacturers who were worried about a single machine dominating the market.

"The meeting with manufacturers led nowhere. Some wanted the DTI to pay the costs of developing compatible BASIC interpreters, which was a non-starter. What we'd failed to appreciate until then was that only two or three had the ability to write ABC. Meanwhile, Newbury struggled to produce a machine, and their progress was very slow.

"Basing the project on existing computers in UK homes would have been a problem. Very few families had computers and those that did usually had a Tandy TRS-80, Apple II, a Commodore PET or VIC-20. The few British computers available in the high street and through mail order had little in common. The alternative was to promote Britain's leadership in computing using an American machine. That would be a PR disaster.

"The project team decided that a new computer was needed to match the project's requirements. They wanted a machine that supported a suite of capabilities not found on anything available at the time, running a version of BASIC that didn't yet exist – and it had to be ready by the end of 1981. There was debate about how to approach this. BBC Education had experience producing programmes, supported by books and learning materials that had been well accepted within education. Mass-producing computers was another thing entirely."

My role in the selection process

"The clock was ticking. We had a target broadcast date, but nothing to present. John Coll and I wrote an outline of what was needed. Just before Christmas 1980, John Radcliffe wrote to seven UK manufacturers asking whether they could provide a computer able to meet our list of requirements. It wasn't a full specification, just a list of features we would like to see. These included colour graphics, sound, analogue-to-digital converter, rugged packaging with a positive keyboard, and a well-structured BASIC (like ABC). To meet the needs of BBC R&D, it should also have a Teletext display and possibly a decoder. It should be able to integrate into the BBC's studio systems, with outputs for high-quality monitors and domestic television sets. Only Sinclair, Acorn, Tangerine, NASCOM, Transam and Newbury were interested.

"In January, John Coll, Mel Draper (DTI) and I started visiting the six respondents to see what they could offer. John and I visited Tangerine in Ely, followed by Acorn and Sinclair in Cambridge. I remember the day as if it was just yesterday: Tangerine could only offer the CP/M operating system which better suited office applications.

"We visited Acorn later that morning. They wanted to show us what they were working on, but needed more time, so they suggested we come back later in the afternoon.

"We next called in to Sinclair's offices. Clive showed us the flimsy rubber keyboard he planned to use in his new Spectrum. He started with a charm offensive: 'I have WH Smith and all the other big retailers on board, and together with us the world is your oyster.' When that didn't seem to be working, he tried to intimidate us: 'On the other hand, if we don't work together, we'll fight you tooth and nail.' It was an uncomfortable meeting. I recorded his words in red ink in my notebook: '*We'll fight you tooth and nail.*'

"Taken aback by his approach, we returned to Acorn to see a demonstration of what they'd been working on, and we were quite impressed. We finished the day at a local pub with a few of their people. What we learnt from the time we spent with the Acorn engineers was that they were clearly very capable, confident of their own abilities, and the sort of people we could work with.

"John Radcliffe chaired a selection committee meeting, to which representatives of the DTI and BBC R&D also attended. We discussed the merits of each proposal and whittled the list down to three companies. Acorn was one of them, but not the preferred option. Newbury was still in the running but only because of its relationship with government. Tangerine was seen as the best choice, though CP/M was not the ideal operating system.

"In early February, when Steve Furber, Sophie Wilson and others in the cramped office at 4A Market Hill managed to build the Proton in four days, it convinced us that here were the people we needed to get the job done. In the final analysis, that influenced our decision.

"The difference between Sinclair's and Acorn's approaches influenced the arguments a week later when judging the proposals from the various suppliers. Sinclair had been difficult, but Acorn was willing to work with us,. They saw the advantage in working with our R&D people. In the late January meeting, John Coll had criticised Acorn due to the Atom's limitations, but after seeing the Proton, he argued in favour of Acorn Computers at the final meeting.

"I remember clearly the day we chose Acorn as the preferred supplier: Newbury arrived with its machine, only to find that it still didn't work. That was profoundly embarrassing and disappointing for them but they had run out of time to show what they could do.

"Once Acorn was chosen to produce the BBC Microcomputer, the rush was on to develop it. John Radcliffe, David Kitson and others met regularly with Acorn in Cambridge to discuss progress. My only contribution was to insist that it should include a relay to start and stop the tape recorder that would be used to load programs into the machine. Manufacturing ran into teething troubles in 1981. Acorn struggled to get the machine onto the market so the transmission dates for Paul's TV series had to be pushed back by three months.

"John Coll (with help from Rob Napier) wrote, and I edited the *BBC Microcomputer System User Guide*. I also commissioned software for a *Welcome Pack*. It was a novel idea to provide a series of programs to demonstrate the capabilities of the computer. This was challenging as our timeline forced them to write it before the computer was ready. I also approached people like the Consumer Organisation *WHICH?* to create software. Their *Tax Calc* was one. BBC Enterprises also started thinking about producing software."

Outcomes

"Once Acorn was able to ship machines, we demonstrated BBC Micros and the *Welcome Pack* to the Ministers from the Department of Trade & Industry – Kenneth Baker, John McGregor, Norman Lamont, John Wakeham and Patrick Jenkin. I was amused when Kenneth Baker, by then Minister for Information Technology, patted the machine and said, 'Oh it's a bit like a typewriter.' He was very supportive and showed it by appearing in some television episodes.

"1982 was declared Information Technology Year (*IT82*), which supported educational software development and the Computers in Schools project, where 85% of primary and 65% of secondary schools chose BBC Micros.

"By 1983, the BBC Broadcasting Support Services had answered 300,000 enquiries, providing advice and referral information by post and telephone. The books and software sold well, and the public had a voracious appetite for information about home computing.

"The extraordinary phase-one response prompted a live programme, where we took over the whole of BBC1 for a Sunday morning live phone-in and audience show that we called *Making the Most of the Micro – Live!* A similar programme was broadcast in June 1984. The Controller of BBC2, Graeme MacDonald, liked the idea of a live series and commissioned several years of regular broadcasts called *MicroLive*. This series continued to use the machine to explain things like advanced graphics and artificial intelligence – years ahead of its time. One of the last in the series included a demonstration of the Acorn RISC machine.

"Until 1987, *MicroLive* was a regular magazine programme – winning the *Times Technology Programme of the Year Award* in 1985 and 1986. Other series included: *With a little Help from the Chip, The Learning Machine, Micro File, Electric Avenue* and *Micro Mindstretchers*.

"Paul and I disagreed about the BBC Micro. He was always lukewarm about us having our own machine and was more interested in a theoretical account of computing. I was more interested in a practical course showing things a computer could do, and how to do them. The machine did appear in some of Paul's programmes, but he didn't use it often, which worked out well for the later series where we did just that.

"The second series, *Making the Most of the Micro* achieved higher audience ratings than the first. People wanted to use the machine, which by then was selling well.

"We didn't push the computer, but used it to explain principles, which included programming with BBC BASIC. It was used for graphics, telecommunications, music and control applications; and it performed brilliantly in the studio. The series combined demonstrations, coding and applications from the real world and research. A third series called *Computers in Control*, looked at robotics, where it controlled the BBC Buggy, which was programmed to find its way out of a maze.

"John Radcliffe and I attended the launch of the machine in the USA. Sadly, we also covered Acorn's difficulties when it was sold to Olivetti.

"The decision to select Acorn Computers to manufacture the BBC Microcomputer under licence proved to be a key factor in the success of the project. Its influence didn't end in the 1980s. The Acorn team went on to develop the Acorn RISC Machine using the ARM chip, which powered the next generation.

"The BBC Micro sold by direct mail, through retailers, and the *Computers in Schools scheme.* We had expected to sell 20,000 computers, but achieved more than 50 times that number, with more than a million machines sold in Britain and overseas.

"By late 1983, with the second phase of the project underway, and with most of the early production delays resolved, it could be purchased directly or from high street stores.

"Kenneth Baker also set up the Scottish Microelectronic Development Programme (SMDP) and the Microelectronics in Education Programme (MEP) to fund software development. In April 1981, Richard Fothergill released the MEP's strategy document, which helped frame John Coll's thinking. The BBC's *Guidelines for Software Writers* is still well regarded internationally.

"The MEP and SMDP influenced microelectronics, computer assisted learning, computer studies and information handling. They emphasised regional collaboration to help prepare school children to live in a society where microelectronic devices and systems would be commonplace. How right they were.

"When Peter Hennessey, the constitutional historian, interviewed Kenneth Baker, the Minister explained that Mrs Thatcher was keen to get computing moving; so, among other initiatives, he proposed government funding for computers in schools.

"The PM later wrote to say how she'd followed the project with interest. Senior BBC managers were interested in the machine and supported our choice of Acorn, ignoring objections from Sinclair and others. I presented the machine to the BBC Governors, once it was launched, and discussed the concept of Telesoftware with Mrs Thatcher.

"The star of the show was the BBC Micro. Though some computer suppliers protested, it was brave of the BBC to develop and sell its own hardware and software. It had the potential to be controversial but came to be well received in most circles, especially among viewers.

"Royalties from sale of hardware partly funded TV programmes and the 1985 *Domesday Project* – a nationwide interactive database of maps, photographs, reminiscences and news that captured life in that year.

"The way I view it: Computers came out of the hands of the few and into the homes, schools and high streets in the early 1980s.

"The second phase of the BBC's *Computer Literacy Project* ran from 1983 to 1989, changing its viewing audience in ways that are still being felt today. It inspired a generation of coders and led to a wave of new educational software.

"Though PCs, Macs and the Raspberry Pi have replaced most of them, BBC micros and their offspring can still be found in some classrooms and laboratories doing what they were designed to do.

"Finally, let me share some what-if questions that I've asked myself many times over the years:

What if the Thatcher Government came to power a year earlier?

What if Acorn hadn't been awarded the contract?

What if I didn't see the possibilities of becoming involved with a subject about which I knew almost nothing?

Well, the stars did align; and the effect of those and other events impacted the digital world we live in today."

BBC Computer Literacy Project Archive

"It's all very well having the BBC Micro for the TV broadcasts, but without software, it might just as well be used as a plant stand. Most of the demonstration software used for them was written by a brilliant studio engineer called Steve Lowry. The programs were explained by John Coll while he was quizzed by Ian McNaught-Davis – 'Mac'; but the behind-the-scenes magic was the result of Steve's hard work.

"Steve contributed to the project for many years. After retiring, he collated the material from 10-years of programmes into a BBC-hosted archive website which also contains his studio software, running on a BBC Basic emulator, and much more. It's well worth a visit."

00000011

The Langham Decision

When the BBC began searching for a UK computer manufacturer, Acorn Computers operated in an environment where it had no serious local competitors. In 1980, the market was vastly different from what it is today. US companies Commodore, Apple and Tandy dominated home computer sales. Sinclair had the ZX80, but like most British home computers, it was little more than a toy.

The Government was willing to invest in technology, which presented an opportunity for UK companies. Of those with any home computer design and manufacturing experience, Acorn had the most developed machine (the Atom) and the most developed BASIC programming language. They had another machine based on the Eurocard format – circuit boards fitted into a backplane. While this as-built design was not in contention, it did mean they had experience and proven designs for the add-ons they'd need to produce, such as disk drives, video displays and Prestel. This all evolved through strong technical leadership of its engineering and design team. Most were Cambridge graduates.

The University of Cambridge was catchment for some of Britain's best and brightest. Hermann Hauser and Andy Hopper could recruit from a stellar selection of hardware and software engineers. The University encouraged staff to develop other interests and were free to share information and resources. Seldom seen at other universities around the world, that freedom to act benefited Cambridge and the entire country.

Among Acorn's alumni are some who didn't study at Cambridge but shared Hermann's and Christopher Curry's vision and confidence. That sort of leadership drew like-minded souls from as far away as the USA, Canada, Australia, and New Zealand. They contributed to its success too.

One of interesting aspects of this story is how and why the British Broadcasting Corporation chose a small business that was operating from a few offices behind a retail shopfront in central Cambridge to produce the BBC Microcomputer System.

BBC Tender Evaluation

There were six fledgling UK microelectronics businesses to choose from. Considering how small all those companies were in 1980; it seems unlikely that any of them would have been awarded the contract if the BBC had known they would be making and supporting more than one million machines – but the world was a very different place back then.

To understand how this happened, we need to go back to 1976 when the Government's National Enterprise Board (NEB) bailed out Sinclair Radionics. Two years later, it provided additional funding for Sinclair to develop a computer that could compete against the Apple II.

After Clive Sinclair left to join Christopher Curry at Science of Cambridge in 1979, the NEB sold off the Sinclair Radionics calculator business. The computer didn't have much to show for two years' work, so the buyer wasn't interested in it. Early in 1980, the NEB hoped to recover its investment by assigning the project to another NEB-assisted company, Newbury Laboratories in Cambridge, which named the machine: the *NewBrain*.

One thing a government-owned organisation understands is another government-owned organisation, so it's not surprising that Newbury was the British Broadcasting Corporation's initial choice to build the computer. It was to be based on the *NewBrain*.

The meeting called by David Allen and Ray Kernow in the autumn of 1979 failed to interest UK companies. They wouldn't work together to resist foreign competition by developing ABC – *Adopted Basic for Computers*. This was when Christopher Curry learnt that the BBC planned to promote computing in Britain. By late 1980, Newbury was getting nowhere. It was increasingly unlikely that Newbury could deliver anything like the machine that the BBC needed to start broadcasting, so urgent and effective action was needed.

It was difficult to find records to prove that the tender selection process was fair. The investigation that fixed the various dates would have made Miss Marple proud. This involved finding supporting documents and speaking to the evaluation team. The goal was to find when the BBC visited Acorn and to learn how they made their decision. Being an open tender, any hint of bias would call the process into doubt.

Spoiler alert: The tender process was completely impartial. What follows is the order of events…

13 December 1980 With the Acorn Atom selling well, it was time to plan for what would come next. Steve Furber wrote 'Essential Specifications' for Acorn's next computer: the Proton. It describes a business computer with integrated monitor and keyboard, disk drives and printer interfaces. It was to be a dual processor machine with a 'channel' connecting them.

16 December 1980 John Coll visited Newbury Laboratories. In his report to John Radcliffe a few days later, he stated that the future of their favoured ABC BASIC was in doubt. If it was to go ahead, funding would be needed – possibly from the DTI. He noted that the graphical interface for the *NewBrain* were very limited and suggested that the NEC be warned to allow for it in its course development. He wrote:

> ...the graphics commands are primitive and may seriously detract from the machine's appeal to the lay audience. Coupled with the fact that the BASIC is not fast and that machine code/assembler work will not be easy for the owner, it seems probable that good games will be difficult to write. Games are an important selling point.
>
> I suggest that the BBC does not adopt the cassette SAVE formats for BASIC programs that are to be used on the Newbury machine. See page 7.
>
> It was only possible to spend about 10 minutes using a Newbury computer. I have no doubt that other things would be discovered if sufficient time was available. It is vital that a thorough review is made available to those who are authoring applications packages and courses associated with the television series.
>
> There is no IF THEN ELSE construct.

17 December 1980 John Coll sent a copy of his report to the Newbury Laboratories CEO, who responded four weeks later.

23 December 1980 Two days before Christmas John Radcliffe wrote to seven British companies seeking expressions of interest to produce a BBC-branded computer under licence. Acorn, Nascom, Newbury, Science of Cambridge (Sinclair), Tangerine, and Transam Components were interested. Research Machines declined.

January 1981 Roy Williams of BBC Merchandising wrote to the six companies outlining the specifications for the BBC Microcomputer.

Each vendor could pitch the best features of their offer. The BBC wanted a machine that supported a highly structured BASIC; a broad range of software; CP/M for small business users; text and graphics for educational and games software applications, and external devices including Prestel/Teletext. The selection criteria focused on the qualities of each supplier: its manufacturing capability; a proven track record; its commercial ties with suppliers; and the ability to work with the BBC.

10 January 1981 A copy remains of Steve Furber's update to his evolving hardware description of the Proton that would guide their work designing it. He doesn't appear to have seen Roy Williams' specification.

11 January 1981 *Technical Appraisal of Contenders for the BBC Micro* – Henry Budgett's report was presented to the February 12th meeting when the final selection was made. It's clear from his comments that he doesn't favour the Acorn Atom but is impressed with Acorn's people:

> The Company: -
>
> Offshoot of the Clive Sinclair empire which founded its relative success on the Atom, a computer they failed to develop to its true potential. Possibly too small to fulfill the estimated demand. They already have the machine built elsewhere. With decent backing the system could be adapted to suit but is likely to meet some resistance due to the shoddy nature of the early product and its existing reputation of non-compatibility.
>
> Personal Comment
>
> The company have been very forthright in their tender for the machine, if they are prepared to do all the work and to support the product then this could fill the need. For further comment see the attached note.

15 January 1981 John Radcliffe wrote a critically important direction to the evaluation team:

> '...sets out the considerations that will determine the selection of the company that will produce the BBC microcomputer'

It listed 14 points addressing technical considerations and what they were looking for in the supplier. It also mentioned the BBC's 10% royalty.

22 January 1981 The CEO of Newbury Laboratories wrote to John Radcliffe. He was responding to John Coll's and Henry Budgett's reports; advising that he would visit his solicitor the following day to discuss licensing. He clearly felt that Newbury was still in contention.

John Coll painted a grim picture of the shortcomings in the Newbury offering. Choosing another supplier was the BBC's only option.

23 January 1981 Site visits were completed on 22nd January, so John Radcliffe called a meeting for 30th January to trim the list of six suppliers.

30 January 1981 The list of suppliers was reduced to three: Acorn Computers, Tangerine Computer Systems, and Newbury Laboratories.

4 February 1981 John Coll wrote to John Radcliffe. He was unhappy with the way the meeting went on 30th January. He discussed the pros and cons of the three remaining contenders. In the case of Acorn, he mixes details about the Atom, the Eurocard Systems and the Proton. From his comments, he appears to know that the Proton is being prototyped at the time, and quotes from Acorn's 'Super Basic' specification.

> The new version of BASIC is very nearly Microsoft but incorporates numerous extensions - some essential, some very useful for the more advanced programmer. The bad areas, in my view, are
>
> (a) Non-standard PRINT format,
>
> (b) non-standard INPUT and INPUT LINE,
>
> (c) Non-standard statement separator,
>
> (d) poorer output formatting of numbers than the Newbury (which looks as if it will be exceptionally nice),
>
> (e) not as conventional file handling as one would like. I am certain that none of the above would be difficult to rectify.
>
> Against these I would set the field proven hardware and software, the demonstrable graphics and two disk operating systems, the extensive range of expansion units which include a P.O. approved PRESTEL unit, the range of available High-level software which includes a good word processing package.
>
> Looking ahead (which everyone else seems to rely on) the dual processor architecture of the Proton is streets ahead of the Newbury and Tangerine machines and will ensure a long-term future for the basic machine.
>
> Contrary to the impression given at the meeting Acorn do have adequate U.K. production facilities and have Open Accounts with semi-conductor manufacturers.
>
> Independently of this report the machines are being tested and assessed...
>
> Acorn – David Allen/John Coll
>
> Newbury have not delivered to me, as promised, a machine nor have they telephoned to offer an explanation.
>
> With Tangerine: ... The hardware is very neat but is totally unsupported by expansion units. They are a company with a very undeveloped product.

6 February 1981 Chris Turner's notebook shows that the BBC visited Acorn. This was the day the Proton prototype operated for the first time.

Some entries in Mel Draper's diary confirm key dates and times as outlined above.

Tu27 Jan81 10:00 - With David Allan to Acorn computers
Fr30 Jan81 09:00 - To BBC all day
Fr06 Feb81 10:00 - John Radcliffe, to Acorn at Bridge Street, Cambridge
Th12 Feb81 14:30 - BBC at Conf room 357, Langham, Portland Place

John Radcliffe, Mel Draper and John Coll arrived at 27 Bridge Street at 10:00am. Hermann and Christopher gave a brief explanation of what was planned; then they all walked to Market Square to view the prototype.

People present on that day report having lunch at the *Baron of Beef* after the presentation. Over lunch, the BBC assessment team questioned Steve and Sophie to get more details about what they were shown.

On that same day, John Radcliffe wrote to Commodore explaining that the machine must be British made but would consider their add-ons.

12 February 1981 The BBC evaluation team met to make its final selection. The prepared agenda was headed with the machine's official title 'BBC Microcomputer' for the first time. The meeting started at 14:30 in Room 357 at the Langham Hotel, which was owned by the BBC at the time. Reports were presented by John Radcliffe (Newbury), Mel Draper (Acorn) and John Coll (Tangerine). They compared the hardware, software, companies and price. The selection panel unanimously agreed that Acorn Computers should produce the BBC Microcomputer.

All contenders were informed of the decision. It appears that Acorn staff may have cracked open the Champagne on Friday, 13th February to celebrate their success.

For the final word on the selection process, who is better qualified to explain how the decision was made than John Radcliffe – who was responsible to the BBC and the Government of the day? John prepared his report in April 1981.

Refer *00011111 – The Radcliffe Report* to see a full copy of the report.

00000100

Richard Russell

Now retired from BBC Engineering, Richard Russell has championed the virtues of BBC BASIC for more than 40 years. A graduate of Hertford College, Oxford, Richard was a junior engineer with the BBC when computer education was introduced into its television programming. While liaising with Acorn Computers on behalf of the BBC, he developed BBC BASIC for other platforms. That became a lifetime interest so that it now runs on most computers available today. This is Richard's story…

"When, towards the end of 1980, it was realized that Newbury Laboratories might fail to meet the BBC's specifications or timescale to design and build a BBC microcomputer, other British computer manufacturers were invited to tender for the work. BBC Engineering was asked to provide technical assistance during the evaluation process.

"Once Acorn was chosen, the BBC maintained a watching brief as the BBC Micro took shape. David Kitson, Head of Transmission Group, Designs Department, delegated much of the day-to-day activities to me, as I had some experience with microprocessors; though that experience was mostly with the Z80, not the 6502 that Acorn planned to use.

"One of my first responsibilities was to draft the specifications for the BBC/Acorn Computers supply agreement. In July 1981, David and I made one of many visits to Acorn in Cambridge. This one was to check on hardware developments as the first prototype circuit board was being built and tested. We were interested in the numbers and types of chips needed and tried to anticipate potential supply problems. We doubted Acorn's confidence that the Ferranti uncommitted logic arrays (ULA) would be ready in time. Our concerns proved to be well founded.

"I also acted as a point of contact for Acorn when they needed help from our engineers, and to ensure that the BBC's technical requirements were met. This included liaising with Sophie on the design of BBC BASIC, which evolved as a compromise between what she was planning to produce for the Proton and the BBC's desire for a language that was largely compatible with Microsoft BASIC, but with features that supported structured programming."

BBC BASIC Development

"When I became involved in the selection of Acorn as supplier and in the specification of BBC BASIC, I was a fairly proficient programmer of Z80 assembly language, in my work for the BBC and as a hobby. My interest in interpreted programming languages was piqued by my involvement in the project, and I realized that I understood enough of the inner workings of interpreters to write a version of BBC BASIC for the Z80, although at that stage it was just for my own education and amusement.

"In June 1981, I wrote a footnote to a business letter to Hermann Hauser that stated my intention to write a Z80 version of BBC BASIC in my own time and asked for some 'inside information'. I must have received help, as some parts of my versions have adopted Sophie's approaches.

"When completed, I realized there might be a market for my BBC BASIC (Z80). It first went on sale in 1983 for the CP/M operating system, which ran on 8080 and Z80 machines. CP/M didn't have a standard graphical display, so it was limited to text output.

"With the success of the BBC Micro and the *Computer Literacy Project*, other companies wanted to offer BBC BASIC for their machines. Since, many of these computers were based on the Z80, I only needed to adapt the graphics, sound and other unique features of each machine.

"Between 1983 and 1986, I created versions for the Torch Second Processor, Acorn Z80 Second Processor, Wren Executive, Tatung Einstein, Research Machines RML 480Z and Amstrad CPC 664/6128. Unlike the BBC Micro, which held BBC BASIC in ROM, the interpreter for these had to be loaded from a floppy disk. Users didn't mind as it offered them better programming tools.

"By the mid-1980s, sales of IBM PCs and compatible machines had taken the lead in home and business computing, so I received requests to implement BBC BASIC for *Big Blue*.

"The first IBM PCs were fitted with the Intel 8088, later replaced by the 80x86 and Pentium families. It wasn't fully compatible with the Z80, but the 8088 and the Z80 had both been developed from the earlier 8080 processor, so they had many similarities. Converting my code was easier than if I'd been starting from scratch. Two friends at the BBC, Brandon Butterworth and Jeff Raynor, translated much of the generic code while I concentrated on hardware and operating system-specific features, like disk storage, graphics and sound. The IBM PC version of BBC BASIC was released in 1986.

"Research Machines had been a market leader in educational

computing before Acorn Computers. They realized that BBC BASIC and through it, compatibility with the BBC Micro, was a way to hold off the threat from IBM PCs in schools. Though they were not IBM compatible, RML Nimbus computers were based on the 80186 (from the same family as the 8088) and used MS-DOS as the operating system. Research Machines adopted a hybrid solution that emulated the BBC Micro's graphics, sound and operating system for use with my interpreter.

"The Z80 version still had some life left in it. In 1987 Cambridge Computer (a Sinclair company) released the Z88 notebook computer with my version of BBC BASIC in PROM; so finally, those two old adversaries – Acorn and Sinclair – shared something in common. The floating-point arithmetic package in BBC BASIC was also used by the Z88's resident spreadsheet/word processor application, PipeDream.

"Four years later, Amstrad released its Notepad computers – the NC100, NC150 and later the NC200. They also used my version of BBC BASIC (Z80) as the resident programming language in ROM.

"Pentium PCs running Microsoft Windows could run programs more than 100 times faster than the original 8088 processor, so I considered that there was little need for further development of BBC BASIC. Mercifully no serious bugs were found; and apart from a couple of minor updates, it remained substantially unchanged. Although it wasn't fully integrated into the Windows environment, it worked well in a window, where BASIC programs could even be controlled by the mouse.

"In 2000, I was persuaded that a native Windows version would be worthwhile. I had no idea, when I began, whether I had the ability or the stamina to produce a version good enough to sell; but if I'd known it would take 18 months, I might never have started. However, my efforts paid off: BBC BASIC for Windows version 1.00a was released in 2001.

"In 2015, the Simple DirectMedia abstraction layer (SDL) was released, allowing me to port BBC BASIC to run on Mac OS-X, Linux, Raspberry Pi and other platforms without much modification. I later created SDL-based versions for Android, iOS and for browsers. That required a rewrite in C, which was a major effort that took several months to complete.

"Over the years, I improved BBC BASIC through language extensions. The challenge has been to make it competitive with modern languages like Python while maintaining compatibility and ensuring that changes are made in the spirit of the original.

"I've also added support for features like proportional spacing of fonts, Unicode and right-to-left scripts, anti-aliased graphics, 3D graphics and physics simulations.

"BBC BASIC today is far more sophisticated than it was in the 1980s, yet it remains compatible with those earlier versions. There are countless examples of its versatility. For example, as the world switched from black and white to colour television back in the 1960s, the BBC made black and white copies of many colour recordings. Some of the colour versions were subsequently lost but the black and white copies remained. It is possible to retrieve the hidden colour element from the black and white recordings using a program I wrote in BBC BASIC.

"It has also been used to monitor earth tremors at the Colosseum in Rome, and to track cameras in real-time in a Virtual TV Studio.

"When I was chosen to help the BBC evaluate Acorn Computers and to help specify BBC BASIC, I had no idea that my association would continue throughout my career. I recall being disappointed that Acorn would base the BBC Micro on the 6502 – a processor I knew little about. However, the source of that disappointment presented an opportunity. I think there is a moral in there somewhere."

00000101

Sinclair v BBC v Acorn

The *Computer Literacy Project* began for all the right reasons, but it was a turbulent era in post-war Britain. It was in that time of rapid change that the BBC Microcomputer was born. Market research is only as reliable as the questions asked and the ability to analyse the data. There is no shame in acknowledging that some assumptions were completely wrong; but even BBC internal correspondence avoids addressing its own contribution to the problems created by its runaway success.

In Cambridge, Clive Sinclair, Christopher Curry and Hermann Hauser had a lot in common. Even the BBC's first attempt to develop a computer began with a machine from Newbury Laboratories that was first conceived at Sinclair Radionics. Let's look at some of the BBC's decisions and its relationships with Acorn and Sinclair during the turmoil of the 1980s.

The Sinclair skirmishes

In January 1981, BBC Television producer David Allen met with Clive Sinclair at his offices in Cambridge. He noted that Clive promised to 'fight tooth and nail' if Science of Cambridge wasn't chosen to produce the BBC Microcomputer. He meant what he said.

Will no one rid me of this turbulent priest? The question Henry II is said to have asked might have been just as relevant for the British Broadcasting Corporation 800 years later. Like Thomas Becket, Clive Sinclair rejected anything that didn't fit his world view. After the BBC selected Acorn Computers as project partner, Clive mounted a one-man media attack, uttering anything that would get him column-inches in UK tabloids. His influence with members of the House of Commons led to occasional questions being asked, though he never received any serious support.

John Radcliffe and David Allen of Continuing Education Television, and David Kitson, Head of Transmission Group, Designs Department, stood by their decision. Far from damaging Sinclair's business they believed he benefited from growing public interest in computing.

Within the BBC, Clive was described as 'single-minded to the point of obsessiveness, with ambitions to dominate the global microcomputer market'. His support came from 'a few journalists who saw the chance to make headlines by fanning his obsessions and his impressive disregard for anything that didn't support his position'. Clive's comments were not shared by Whitehall, educators or the rest of the UK computer industry.

In July 1982, more than a year after the BBC chose Acorn Computers to develop the BBC Micro, Clive Sinclair came to call. Ignoring him hadn't worked. Senior management needed to find a way to get rid of him once and for all. He was never one to let the facts get in the way of a good story, but the facts offered them the means to deal with him.

The *Computer Literacy Project* team engaged in strategic and tactical planning, not seen since the Queen's coronation. They scanned press statements to understand his messaging and built their defences to respond head-on to some of his more ridiculous claims.

Misstatements were Clive's stock in trade. For example, he once claimed that a ZX80 microcomputer could run a power station! They knew they couldn't reason with him; and clearly, a frontal attack was not the answer.

While Clive Sinclair was not one to be trifled with, he wasn't a planner. He just shot from the hip. The BBC was going to war, and its adversary was hopelessly outgunned for the exchange that was about to begin. The senior managers who attended the meeting were well briefed. Each received summary points, short lists of the main issues that Sinclair seemed likely to raise, as well as longer, more detailed briefing notes.

Unwilling to give him an inch, John Radcliffe believed that if they made any concessions, Sinclair would be back for more, so the meeting aimed to silence him. They prepared to do what the BBC does best: It presented a strong, coordinated stream of irrefutable facts. Responding to his claims and demands, they met his attacks head-on with well-prepared, well-rehearsed and well-considered responses that blunted his offensive.

Bruised and beaten, little more was heard from Clive after that meeting as he focused on developing new machines for the consumer market – until the BBC Microcomputer contract was due for renewal in 1984/1985. It may seem hard to imagine now, but a Sinclair BBC Microcomputer was once a serious possibility.

Relations between the BBC and Acorn Computers

In the first few months of 1982, after several frustrating delays and setbacks, production started in earnest at ICL's Kidsgrove plant and then at Cleartone in Wales. About 13,000 machines had been shipped, with orders in hand for 12,000 more. By early May, Roy Williams (Head of Merchandising for BBC Enterprises), David Kitson and John Radcliffe had prepared a no-holds-barred review – *Relations between the BBC and Acorn Computers*. It was a critical but positive plan that proposed practical steps to help Acorn and to protect the BBC's reputation.

Roy Williams was responsible for day-to-day commercial aspects of the contract. David Kitson monitored development and production. John Radcliffe played a co-ordinating role. Along with BBC Publications, they held weekly progress meetings with Acorn. While they acknowledged Acorn's technical abilities, they were concerned about its performance.

David Allen raised his concerns in another report. Both acknowledged the difficulties Acorn experienced in meeting demand for product and adequate customer service. They agreed that the cause had been:

> "... unforeseen technical problems in achieving mass production of a new and exceptionally advanced microcomputer, and partly because of managerial weakness within Acorn, a small company which has had to grow relatively fast; and manage a manufacturing operation on a scale and of a complexity beyond its experience; all this in the glare of publicity which has surrounded this project. BBC staff have provided a steady flow of advice and help to Acorn, tempered by what little pressure it has been possible to exert under the terms of our contract with them. Recent difficulties suggest that the time may have come to modify some of the terms of the contract and require some structural changes within the management of the company, in order to achieve more effective provision of this key element in the Computer Literacy Project. It should, however, be borne in mind that despite all the problems which have arisen, the machine itself has attracted extremely favourable comment from nearly everyone who has assessed it."

Despite Acorn's confident prediction that production would be running at 3,000 a week, ICL was producing only 800 Model B machines, and Cleartone was building 600 Model A. New orders had dropped from 1,000 to around 300 a week. They blamed an apparent drop in consumer interest on bad publicity and the BBC's refusal to allow Acorn to advertise while the backlog was so large. Considering that the original agreement between the BBC and Acorn refers to what should happen if total sales failed to reach 12,000 units, it seems that no one anticipated how successful it could be. It is hard to imagine how Acorn managed to ship one million machines, but serial numbers known to have been produced support that number. Does this mean that the BBC's advice helped Acorn overcome its growing pains? The report stated:

"We have constantly pressed Acorn to increase the level of production at both subcontractors, and to contract a third manufacturing source. These efforts are being endangered:

(a) by an apparent unwillingness on the part of ICL to give high priority to this project (ICL may not be prepared to commit themselves to a further contract, which has been offered to them by Acorn),

(b) by the financial weakness of the Cleartone Group, which is currently in the hands of the receiver,

(c) by delays in finding a third manufacturing source (currently at least 12 weeks away from production), and

(d) by an apparent general difficulty on the part of British electronic subcontractors to meet their promises and hit their targets.

It should be said, however, that this is not, technically, a particularly difficult machine to make."

The BBC was beginning to see just how backward British electronics manufacturing had become. They asked the Department of Trade & Industry to put pressure on ICL to achieve more reliable deliveries. In a remarkable shift from earlier policy, they proposed that Acorn might engage an Asian subcontractor to help meet the backlog.

The failure rate of new machines was at least 10%, blaming teething troubles in production and inadequate quality control as the causes. To reduce failures, and to protect the BBC's reputation, they proposed appointing two BBC Engineers full time to monitor production.

Acorn had subcontracted distribution to BL Marketing (a subsidiary of breakfast food giant, Weetabix), but erratic deliveries from ICL and Cleartone, and weaknesses in BL's record system had been a formula for failure. As a result, the Weetabix Group decided to shut down BL Marketing. The BBC supported Acorn's decision to set up a distribution arm, and to recruit BL's former staff, to run the operation under Acorn's direct control, while bringing distribution and information services under Acorn's customer relations and marketing management.

The BBC's Broadcasting Support Service dealt with over 100,000 enquiries in six months. There had been a steady flow of technical questions, enquiries and complaints about shipping delays, many of which were dealt with by BBC Enterprises. It was their policy to deal directly with complaints "in the interests of the Corporation's good name", but they expected Acorn to provide adequate technical information and after-sales service, while the BBC kept the power to control what was said and published.

They believed that the flood of complaints had dropped as production increased but still doubted Acorn and BL Marketing could deal with the workload. They hoped Acorn's new dealer network might help but expected to receive more complex and more demanding enquiries as disk storage, second processors and teletext were released. To manage this complexity, they wanted to set up a central information system under new distribution arrangements.

There were parts of the report that had a pensive tone. They resented the way Acorn handled development of the teletext receiver, second processors and networking, while continuing with Torch and Electron development. They complained that it placed heavy demands on BBC Designs Department staff – "essentially because of the continuing failure on the part of Acorn to set targets and stick to them". The report reflected a view that Acorn should focus entirely on BBC priorities, to the exclusion of everything else.

They believed that Acorn should beef-up its management team.

> "Acorn's top management is overly optimistic, unable to identify and concentrate on essentials, a lack of direction, appalling internal communications and a willingness to carry on the development phase of a product long after it should have called a halt. In this they are probably not untypical of this industry; indeed, Clive Sinclair undoubtedly shares many of the same characteristics, but it does not make them very easy partners for the Corporation. The inescapable conclusion from the analysis above is that drastic improvements in Acorn's management structure is needed. There appears to be a need:
>
> (a) for a Product Manager, responsible within the company for the development and production of the BBC Microcomputer System, with the authority and resources to take decisions and get results,
>
> (b) for a Customer Services Manager, responsible to the Product Manager for distribution, information, and after-sales service.
>
> "Both would need adequate staff support.
>
> "It seems very unlikely that without requiring some very specific changes of this kind, that any real improvement can take place. It is proposed therefore that the BBC should, at its own expense, urgently commission a study of Acorn's management structure, in order to recommend what steps need to be taken, with detailed requirements for supporting staff and lines of responsibility. This would both arm us for discussions with the company, and provide a basis for recruiting the necessary staff, defining their responsibilities, and maintaining surveillance of Acorn's operations, as a continuing service both to the company and to the BBC.

"Even if we are able to meet all or part of the initial cost of making a new corporate plan for Acorn's future operations for the BBC, this will clearly involve the company in extra expenditure. We do not know what their current level of profit is and so are not in a good position accurately to assess how far this may be a problem for them. However, since they have already delivered £M4.5 of equipment, have orders for a further £M5, and could well do a further £M15–20 of business this year, it seems likely that means can be found for financing changes which can only improve the profitability of the company. We do not believe that the initial cost of development (probably less than £M1) and the cost of the technical delays in Dec/Jan, significantly alter this judgement.

"As an alternative to the proposals above it would in theory be possible for the BBC to take on direct responsibility for distributing the microcomputers, for information to customers, or even for production. We believe that to extend our direct responsibilities might make it possible to manage these operations more effectively: it would however be extremely expensive (even if the costs could ultimately be recovered), would take us into operational areas of which we have comparatively little direct experience, might well lead to a dangerous division of responsibility, and would in any case probably take us beyond the terms of the BBC Charter. We suggest that these are not practicable or viable alternatives, and that if we are to continue our partnership with Acorn, we should now bend all our efforts to making it possible for them to carry out their responsibilities under the present contract. The obstacles to this are, in essence managerial, and, with a modest injection of resources now, there is no reason why a managerial solution cannot be found. There is strong support in Whitehall for this view, and if Acorn does not accept our proposals, they are unlikely to be accepted for the Micros in Primary Schools' scheme.

"Under the present contract it might, in theory, be possible to terminate our relationship with Acorn and turn to another manufacturer. However, we do not own the design of the computer, nor have we contributed substantially to its development. The BBC does not, therefore, have the power to subcontract the manufacture of the machine directly. To go for another computer company to produce an alternative BBC machine would be costly, unacceptably time-consuming, and involve a major public row, and might in the end leave us in no better case; the indications are that Acorn is no better or worse managerially than most small companies in this field. We are, to put it bluntly, stuck with Acorn, and have few sanctions that we can impose on them. However, they have a very considerable financial interest in the success of this project, and if, by a combination of diplomacy and pressure we can induce them to accept the steps we feel are now needed, this is in our view, a better option than any more drastic change of course.

"It should perhaps be pointed out that if the present difficulties can be satisfactorily overcome, we will have an excellent base for future developments. The Computer Literacy Project has already achieved its primary aim of giving a vigorous stimulus to computer education in Britain in Information Technology Year. It has been widely acclaimed as an innovative educational project both here and in many overseas countries, and it is a base on which we can build."The BBC Microcomputer System is likely to be widely used in schools, businesses and elsewhere. The software capability being developed by BBC Publications will provide a new and powerful support to broadcasting, initially in the educational field, but probably later well beyond it. The development of a telesoftware service initially linked with this project, is a further new and potentially very valuable development, which underlines the Corporation's world leadership in teletext technology. Using this base, the educational broadcasting departments are developing a range of new series in the field of information technology, which will consolidate and develop the world lead we have established in this field. Lastly, and not unimportantly, these developments are likely to provide the Corporation with substantial revenue, now and in the future. If resolute and skilful action can be taken to deal with the short-term problems we face, the long-term benefits will be very considerable."

The BBC made a concerted effort to encourage and support its partner. For its part, Acorn recruited a strong team of managers. There were several significant changes in manufacturing. The dust settled. Acorn continued to evolve. It raised more capital through an Initial Public Offering on the London Stock Exchange, but consumer interest was changing, and competition was increasing. Acorn would soon face new challenges, the least of which would be renewal of its Agreement with the BBC.

The Sinclair BBC Microcomputer

When it came to personal bias for home computers in the 1980s, UK consumer opinion was heavily polarised between Acorn and Sinclair. By 1984, the shine had worn off the BBC Micro so something new and different was needed. To a devoted Acorn user, the very thought of Sinclair producing the next generation of BBC Micro was heresy, but for a time, it seemed they were in with a chance.

The three-year agreement between the BBC and Acorn Computers took effect in August 1981, with a few amendments made in late 1983. By May 1984, with only a few months left for the contract to run, John Radcliffe, David Kitson and Richard Russell prepared a report addressing possible options for the future of the BBC Microcomputer.

It was agreed that the *Computer Literacy Project* should continue. The computer had played an important educational role, and with substantial commitment by the BBC and other agencies to produce educational software in BBC BASIC, it was believed that a BBC Microcomputer should be produced for at least another two years.

The growing choices in home and office computers made the BBC Micro seem outdated, though most of the arguments were easy to refute. To move with the times, the BBC wanted to licence an updated machine from Acorn, or one produced by another British company. Because of widespread support for the Acorn machine, there would need to be a compelling reason for changing to another company. The working group spent three months examining several options.

Five companies expressed an interest in supplying the next generation BBC Microcomputer: Acorn, Amstrad, Commodore, Research Machines and Sinclair. Again, Commodore was ruled out because it was not British, and Amstrad because it products were produced in Asia. Research Machines submitted a proposal but with no detail. Acorn and Sinclair submitted full proposals for new machines. Theirs were the only offers given serious consideration.

Acorn proposed the BBC Model B+ and Sinclair offered a modified Quantum Leap (QL). The differences in their proposals were hardware and software compatibility and price, as well as concerns about the two companies, their positions in the development cycle, and the BBC's confidence in them.

Acorn had a clear advantage with compatibility: Being a direct successor to the Model B, it would be less disruptive to the software and peripheral suppliers. Schools and other institutions should be able to use the two systems side by side. The Model B+ offered more memory and improved file handling than the Model B. This was an interim solution as the BBC Master would replace the Model B+ a year later.

It's better the devil you know... It had taken the BBC three years to get on top of Acorn's production issues. Changing suppliers at this point would mean starting over. While Sinclair had shipped large quantities of products including computers, they were far less complex, and the level of support provided was well below what BBC customers expected. The Corporation understood Acorn's strengths, weaknesses and dependence on sales of their machine, so it felt that it was in a stronger position to influence Acorn's actions; and besides, with relatively minor design changes needed for the Model B+, they were confident it would be ready by the second quarter of 1985. This was not the case with Sinclair.

The BBC believed that the price of Acorn's new machine was far too high and would be uncompetitive in an increasingly crowded home computer market. In hindsight, price was a problem, but the BBC never lowered its expectations of what the machine should include as standard features, and the level of customer support expected. As Acorn's Chief Engineer, Chris Turner, recalls: "The BBC wanted computing brought to the masses with a high level of local handholding. We realized how reliant users would be on dealers with shopfronts who, of course, needed to make a profit. Distribution could never work as direct mail order only."

The BBC Micro offered many more features than Sinclair, Commodore, Apple, Tandy and other home computers of the time; so of course, the price would always be higher. The Acorn Electron could have solved the problem of entry-level pricing, but not if it had all the extras of the BBC Micro. While it supported some educational applications, the Electron was designed to compete with games machines, where local support was less important. Without the BBC Micro's Mode 7 video, the Electron was not compatible with the BBC's priorities: educational programs, Teletext and telesoftware. This omission avoided paying royalties to the BBC and allowed Acorn to compete against Sinclair and still make a profit.

The report suggested that the BBC should take direct action to influence Acorn's success:

> "We suspect that thus far, despite its technical excellence, the BBC system has sold largely on the name of the Corporation. In our view the apparently unrealistic pricing of the system reflects the lack of an effective Acorn marketing strategy. If the Acorn option is to be acceptable, we feel that the price should be reduced, and that a pricing and marketing plan for the whole new system, including peripherals, should be presented to the BBC."

If the BBC's view of its market influence was realistic, it might have had a big impact on sales of a BBC Electron. It's difficult to imagine how any marketing plan could have lowered the price of the BBC Micro with such high UK manufacturing costs, and all the features that came as standard; besides, despite complaints about the price, most sales were for the more expensive Model B.

The report also forecast the coming of the BBC Master in 1986, without stating what it would be. As always, they were talking up the need for more functions and features – and with that – even greater price hikes:

"For 1986 and beyond it would be necessary to consider, at an early date, the longer-term development of the system. This could include either a highly integrated version of the enhanced machine (very compatible, but with fewer chips, so as to reduce the price), a new model, based on a more powerful processor (which would be less compatible), or both. If the Acorn option is accepted, decisions on these longer-term developments will be urgently needed during this summer. It will in our view be necessary to provide for early discussions with the Corporation on the most desirable specifications for the future development of the system during the period of a new contract.

"Provided that the proposed Acorn machine can be sold at a competitive price, and provided that the company can work out a credible long-term development and marketing strategy, this proposal seems to us to be an attractive option for the Corporation."

But then again ... The Sinclair proposal was based on a modified Quantum Leap (QL). The report described it as:

"...an impressive specification at a very competitive price. It makes rather more efficient use of a larger memory. The design is based on a 16-bit processor, which is inherently more powerful than the 8-bit technology on which the Acorn systems are based. Sinclair have given firm undertakings to meet the details of our specification, and to hold to their quoted prices, although they regard production by the first quarter of 1985 as risky."

Sinclair's Managing Director, Nigel Searle, quoted £299 plus 'extra' for some standard Acorn features. They would later discover that the QL's performance didn't match their expectations for an inherently more powerful 16-bit processor. Despite the QL's 7.5 MHz 68008 processor, BYTE magazine described QL BASIC as "very, very slow" – much slower than Acorn's 8-bit BBC BASIC. As for its suitability for the classroom: Students were known to crash-test BBC Micros by standing on them, and their keyboards were tested well beyond the limits of what a Sinclair QL could withstand.

What concerned the team most was lack of compatibility – both hardware and software. Disk drives and printers might work, but dedicated peripherals like teletext adapters would not be transferable between Acorn and Sinclair machines, undermining the telesoftware service. They also questioned whether programs written in BASIC would run on both systems. This created serious problems for BBC Publications and other publishers that had invested heavily in developing software.

Users, particularly in education, expected BBC micros to be compatible. The original design aimed to be expandable and "future proof".

A change of course would undermine confidence. When buying more machines it would force existing BBC users to make a choice between buying non-BBC Acorn equipment and using two incompatible systems. They worried about how this would affect schools and colleges that had standardised on the BBC machine.

It would also disrupt overseas markets, where sales of BBC computer education materials had benefited from their association with the BBC Micro. They were particularly concerned about the effect that dropping Acorn would have on their relationship with the *Indian Computer Literacy Project*, which had just standardised on the BBC Micro.

The working group expressed doubts about Sinclair's ability to manage the project. While it had successfully sold large numbers of computers in the UK and abroad, it was reported to be suffering financial problems, and there were long delays in shipping the QL. Unable to even supply a sample QL for review, there was little confidence in Sinclair's capacity to meet deadlines or provide customer service and technical support. The company failed to demonstrate its manufacturing and quality control capability, so the working group assumed the BBC would have to set up its own systems:

> "Experience with Acorn suggests that to establish effective liaison with a new licensee is likely to be a difficult process, and it is not one that we could realistically contemplate without the special recruitment of at least two new engineers. Because the BBC contract would not be nearly so important to Sinclair as it is to Acorn, our leverage over the company in practice would probably be slight."

Finally, the working group was concerned that Acorn was heavily dependent for its survival on its link with the BBC. To announce…

> "… a switch to Sinclair could conceivably drive Acorn into liquidation. Sinclair, on the other hand, would certainly survive another failure to get a BBC contract. Sir Clive would undoubtedly make a public fuss about it, but his company would be able to stand the shock."

Choosing two competitors' machines could stimulate the use of BBC BASIC and maximise the Corporation's royalty revenues; however, it was likely to confuse consumers and would be harder to administer and more difficult to market. There were still some unresolved trademark issues, and there would be little compatibility between peripherals and software, making it hard to justify having two incompatible BBC Micros; and besides: if Acorn and Sinclair, why not Research Machines and should they also include Amstrad too?

The two-machine option was rejected. Follow-up correspondence between Sinclair and the BBC didn't display the same level of animosity or rancour experienced after the 1981 decision. This may reflect Clive's shift of focus towards electric vehicles, or perhaps he was just mellowing.

The report concluded that rather than accepting the Sinclair proposal as the basis of a new BBC Micro, it would be preferable to have no BBC machine at all. That would mean abandoning its audience while the educational arguments for having a machine still stood. Many disadvantages of the Sinclair proposal would still apply, and though it might save the BBC from some embarrassment, it was sure to create others. "It would also deny the Corporation a good deal of revenue."

But the report didn't leave it there...

> "However, unless Acorn is able to convince the Corporation that there will be a continuing demand for their enhanced model, as priced, we cannot recommend an extension of the present contract beyond the lifetime of the present system. In these circumstances we believe it will be preferable to abandon the idea of going for a new generation machine rather than accepting the Sinclair bid."

The report made three recommendations...

> "WE RECOMMEND that the bid from SINCLAIR RESEARCH should be rejected on the grounds of its inadequate compatibility with the present BBC system.
>
> "WE RECOMMEND that the ACORN proposal should be accepted, but only if the company can provide more realistic undertakings on price within a properly worked out marketing strategy for the future. This needs to include a long-term plan for the development of the system in 1986 and beyond, which is acceptable to the BBC.
>
> "WE RECOMMEND that if more realistic price commitments cannot be achieved, the present contract with ACORN should be extended, but only for such time as there is likely to be a demand for the present system."

What happened next

The BBC Microcomputer Model B+ was released in mid-1985, serving as a transition between the BBC Models A/B and the later BBC Master series. It came in two variants: The BBC Model B+ 64K introduced additional RAM and modest hardware improvements. The BBC Model B+ 128K featured expanded sideways RAM and support for loading ROM images into memory.

It wasn't a commercial success, partly due to its limited compatibility with some older software and peripherals, but it's an interesting footnote in Acorn's product evolution.

Six months later, the BBC Master 128 was announced in January 1986. It introduced several enhancements, including more RAM, a numeric keypad, ROM cartridge slots, and built-in disk support. Acorn offered variants like the Master Turbo, Master 512, and Master Compact, each tailored to different markets – from education to business to home users.

Acorn machines carrying the BBC logo were sold from 1982 through to 1993, by which time Acorn had shifted its focus to the BBC Archimedes – the last machine to display the BBC brand.

Production of the BBC Master quietly came to an end, yet it lingered in classrooms and labs well into the 1990s – a testament to the durability and the strength of the BBC educational ecosystem.

00000110

Christopher Curry – before Acorn

The story of Acorn Computers begins with another Cambridge business: Sinclair Computers. Hermann Hauser and Christopher Curry enlisted the aid of Chris Turner to develop the MK14 microcomputer evaluation kit for Science of Cambridge – a Sinclair business. This set the pattern for the way they would recruit some of Britain's best and brightest.

Christopher's involvement with Clive Sinclair began in the late 1960s. The relationship changed over the years: starting as an employee, he became a business partner, then later a strong competitor, but he always regarded Clive warmly. Considering the battles they fought in the media, in Parliament, with the BBC and at a pub in Cambridge, it made for an odd friendship. Christopher Curry shows himself to be a loyal, highly creative and an enigmatic leader who, with Hermann Hauser and Andy Hopper, helped to inspire a generation. Here is his story…

"Rather than going to university, I decided to study part time while working at Pye Telecom in Cambridge – a world leader in mobile police radio systems. I started as an apprentice, working on a production line with about 500 women. Those were the days when women employed in UK manufacturing had big hairdos and attitudes to match. It was an extraordinary experience, but I found the work too repetitive for a young chap ready for some excitement, so I left after a few months.

"I saw a job advertised at the Royal Radar Establishment as a lab assistant. They were happy to give me day leave to study at the Royal College of Radar. It was an Aladdin's cave of wonderful scientific things. A beautiful place. Marvellous. Things like the nose cone of the TSR-2 (Britain's answer to the F-111) sat in the lab next to me. It was the most advanced radar in the world at that time. I was making superconducting junctions by sputtering bits of aluminium onto glass. Superconductivity is magic, and great fun.

"Being young and impetuous, I soon found myself in a spot of trouble with management. Let's face it: Any young man who wasn't in some sort of trouble in the 1960s just wasn't trying! I had 'borrowed' a bottle of ethanol absolute from the store to top up my petrol tank for a trip to see *The Beatles* perform at *The Cavern Club* in Liverpool. This wasn't serious – everyone did it to some degree – but getting caught was frowned on, as it created so much paperwork. They also learnt that I'd been skipping classes while on study leave. Youthful rebellion was not a new phenomenon, but it was frowned on by managers who took the defence of Britain against its Cold War adversaries more seriously than I did. To avoid disciplinary action, I chose to leave quietly. I wasn't as disappointed by the forced departure as I was when I arrived at *The Cavern Club* to find that *The Beatles* had cancelled.

"I moved to WG Grace working as a lab technician/taste tester when the company was developing replacements for cork seals in bottles. I tested tonic water samples to decide which one tasted best and closest to the reference; but that was rather tedious too. In 1966, Clive Sinclair rescued me when I joined Sinclair Radionics."

The Sinclair Years

"Clive Sinclair set up his first company in 1961. He had just moved to Cambridge and was recruiting engineers when I met him. Clive was self-taught, so he was quite casual about what constituted an engineer and didn't specify what experience was needed. Mine was building things with valves. As a youngster, I made a few quid building guitar amplifiers using parts from old TV sets from the Council tip. I once built one of Clive's amplifiers, which never worked very well – a feature of a lot of his equipment. In any event, I knew his products and liked building electronic things, so that's where I went. At last I was having fun!

"I developed the Micromatic matchbox radio and some small amplifiers. Clive would come in and say, 'Try this. Try that.' Then he'd disappear. I would build it, fiddle with component values, and look at ways to make it work better. That was my role in the early days.

"Working for Clive was challenging as he was prone to temper tantrums. He would often charge about, shouting and throwing his slide rule around. But when things were calm, it was great fun. We worked late most nights. After a long and busy day, we usually went off to the pub. This was where we talked about new projects – new products to be developed. We'd take Clive's ideas, produce a printed circuit board, populate and test it; then we'd put it all together in a kit suitable for selling by mail order. He always did the advertising.

"New Year's Eve parties were one of Clive's big annual events, and 1971 broke all records for excessive behaviour – not Clive though, perish the thought! I had done my share of drinking and socialising that evening, even though I had to fly to Texas early next morning – hangover and all.

"Clive had arranged for me to collect first samples of the new Texas Instruments single-chip calculator – the TMS 1802. Chips were quite new in those days, usually with very low yields. He had convinced TI to give us three chips from their first five samples. I believe the other two went to Japan. My job, when I got back, was to build a prototype around this chip. We didn't have keyboards or anything to work with, so we made one that used pieces of wire and tin, a clock and an LED display.

"I spent a day or two assembling the parts, but it didn't work on my first attempt; then I found that the clock was set to the wrong speed. One of the most memorable moments in my life was seeing it work for the first time – connecting batteries, seeing the display light up, and lights flashing across the screen; then the seven-segment numbers appeared. I entered a number, divide, another number, then pressed equal to get the result. It was the test we always performed to test calculators: 22 / 7 =.

"When an approximation for *pi* (3.1428571) appeared on the eight-digit display, people gathered around and cheered. This really was magic!

"Sinclair Radionics dominated the electronic calculator market in the UK at the time. Clive claimed that he made the world's first pocket calculator, but that wasn't true. Rapid Data from Canada beat us to the punch with the Rapidman 800, in February 1972, but it was an unattractive green thing. Our Sinclair Executive calculator was released in September. We set the trend, making calculators that were smaller and slimmer than the competition – more compact at a lower price. That's why we were so successful in that market.

"Clive liked to make things look sleek. The pocket-sized case was designed to be a fashion statement – something that wealthier businessmen could show off. In fact, we sold one of the very early models in Cambridge to Lord Rothschild. There hangs a tale…

"We had squeezed all the power we could out of its button batteries – too much in fact, because it caused them to de-gas. His Lordship was in a board meeting, when one of the batteries exploded. Unfortunately, the calculator was in his breast pocket, and he thought he'd had a heart attack. They called an ambulance which found that he wasn't dying, it was the exploding Sinclair Executive. A few days later I took him a replacement. A true gentleman, he was very reasonable about it.

"We changed the circuit after that, so the other batteries no longer tried to charge the one that went flat first. The Executive sold for £79. It cost us £11 to make, and we couldn't make them fast enough.

"There were no managers at Sinclair. Clive supervised the teams, where everyone was responsible for a project – from ordering the components through to getting the item manufactured and packaged up. So, each project was someone's end-to-end responsibility, which was nice, because we learnt every part of the business. Watching him, we learnt how the advertising and marketing were done.

"Clive was ambitious; always prepared to push the limits of good engineering and business practices; but the company was struggling by 1976, due to some serious product reliability issues. The National Enterprise Board (NEB) had been set up by the Department of Trade and Industry under Harold Wilson's Labour Government to bail out struggling UK businesses. Clive was a beneficiary of that largesse, but he hadn't anticipated the level of control they would exert. The NEB invested in Sinclair to help with cash flow problems. Their support came with strings attached: They tried to limit unprofitable activities like the portable television, but Clive wouldn't be hobbled. Always a renegade, he regarded their advice as unwarranted interference.

"The Black Watch and the Pocket Television would prove to be his undoing. The NEB increased its stake in Radionics to 73% when the Microvision hit serious technical problems; then the project was cancelled. While the TV never reached volume production, the Black Watch did – and with it, serious reliability problems. Clive would go for the lowest possible cost of assembly. If he could leave out a transistor that only cost 5p, he would, even if it meant that something might not work quite as well. The lesson from that experience was cut corners on quality at your peril.

"By 1977, Clive decided to distract the NEB team while I helped him ramp up a new business. We used one of his unused companies, Sinclair Instruments, to dispose of surplus calculator chips; and started by changing its name to Science of Cambridge. I borrowed £500 for seed capital from my father to become a minor shareholder; and rented a small office on the top floor of 6 King's Parade, just opposite the main entrance to King's College. Clive moved assets (obsolete components) from Sinclair Radionics. While this wasn't illegal, it didn't show good faith in his dealings with the NEB.

"Using the parts we'd purloined, we released our first product: a Wrist Calculator, based on the C550 chip from General Instrument. It had a lot of buttons and eight small mercury cells. We sold about 100,000 of those watch kits in America, out of reach of the NEB. Selling products as kits using surplus components like the watch calculator was a good way to conceal poor product quality. I don't know how many came back, but there were a lot. People were very patient. They often built them using solder paste instead of a soldering iron, but we could usually repair them.

"Around that time, I was sharing a house with a few university students, including a woman from New Zealand. One day, a young couple came to the house to visit her. It was another New Zealander, Pamela Raspe, and her boyfriend (now her husband) Hermann Hauser. I had known Hermann in the 1960s when he was in Cambridge as a teenager studying English. He was now at the university's Cavendish Laboratory studying for his PhD. After that, we sometimes caught up at local pubs and cafés.

"Hermann and I shared an interest in digital electronics. He told me about the Cambridge University Processor Group and some of what they were doing. It started me thinking that computers for use in the home could be an interesting product for our hobbyist customers.

"I had tried to develop a microcomputer using the C550 calculator chip. Ian Williamson from Cambridge Consultants had been playing with a design for two or three years without success. He then pitched the idea of selling a small, cheap microcomputer kit and I agreed to get involved; but when he was ready to hand over a working prototype, Clive resisted the idea of paying royalties, blaming his change of heart on being unhappy with the design. He refused to sign the contract, leaving me to break the news to Ian. Feeling quite embarrassed by Clive's behaviour, I gave Ian some work to compensate for the time he'd spent on it.

"I had been talking to Sandy Dow of National Semiconductor. Just when Ian's project was unravelling, Sandy offered me a free circuit design plus operating software, in return for an exclusive agreement to use their components. There was even an application note that came with it.

"Though Clive was reluctant, I pressed ahead. The Sinclair MK14 would be a microprocessor evaluation kit – not the sort of thing to reconcile your bank statements or to keep track of your record collection (which were the most popular ideas for using a computer at that time). I just needed to find an engineer who could bring it all together.

"Hermann offered to connect me with people at the university. The first was a young electronics engineer, Chris Turner, who he'd worked with at the Cavendish Laboratory. Chris did a great job with the MK14. My friend David Johnson-Davies wrote some programs to go with it.

They had to fit in 256 bytes of memory so they couldn't do a lot.

"Science of Cambridge released the MK14 in 1977. Its name was inspired by 'microprocessor kit, with fourteen components'. Even though the kit was about as imaginative as its name, it sold by the truckload. It had a hexadecimal keyboard using conductive rubber as the contact mechanism. Sometimes it worked well. Sometimes it didn't work at all. Despite the technical problems, we sold 90,000 MK14s at £39 each. It cost about £19 in parts, so it was quite profitable for us. Clive soon saw there was money to be made selling computers to consumers.

"At first, I was the only person at King's Parade. I placed the ads, packed the watch kits, took them to the post office, repaired some of them, answered technical enquiries – absolutely everything. Later, there was a secretary, then more people were recruited. By the time the MK14 came out I think we crammed about five or six people into that space.

"The MK14 kept me at the office over Christmas in 1977 and 1978. When Santa doesn't deliver, that's serious. We were often late, so I was the elf who had to sort out delivery problems. People wasted no time building it. If it didn't work, they wanted to know why. Imagine it's Christmas Day: lunch is over; the dishes are cleared away; and it's time to assemble the kit. If it doesn't work, they won't wait till the new year get it fixed. That's why I sat in the office through two Christmas seasons answering questions. Most customers were rather nice. There were a few savage ones, but most were very understanding.

"Our market was the hobbyist who liked soldering things together. It wasn't really a *computing* market; it was just an *exploration-of-electronics* market. People liked to send me details about what they had done with it. Using their ideas to improve the product meant R&D cost nothing.

"In our market, the idea of a microcomputer at home was gaining traction. My focus was building computers for the consumer, which I saw as a way to learn programming. While machine code was a starting point, without BASIC or another high-level language, our target market was limited to mathematicians, scientists and engineers.

"We hadn't spotted the business computing opportunity. That came later – though never for Sinclair. There were a few business computers coming from the USA. They were usually built using the S-100 Bus design that started with the MITS Altair 8800, and followed by Cromemco, IMSAI, Morrow, and others. Most of them ran the CP/M operating system and cost considerably more than our customers could afford.

"Clive didn't share my enthusiasm for microcomputers and would need to be dragged kicking and screaming to it – though he later claimed full credit for Sinclair's success. Despite his early protests, the MK14 had proved that we could make money selling computers.

"In 1978, Mike Wakefield and Basil Smith at Sinclair Radionics started working on its first microcomputer project, but Clive left the company in May 1979. The NEB announced the sale of its calculator and TV interests. Newbury Laboratories took over the cancelled computer project. They called it the *NewBrain*. The BBC turned to Newbury when they started looking for someone to develop the BBC Microcomputer.

"Clive took control of Science of Cambridge and brought Jim Westwood along to start work on the ZX80. He changed the company's name again – this time to Sinclair Computers – but that didn't last either. In just eight years, the business had changed its name six times. In 1981, the name changed for the last time to Sinclair Research, reflecting his growing interest in electric cars. Despite all the changes it had produced the Sinclair MK14; then after I left, the ZX80, ZX81, Spectrum and QL series of computers.

"It was time for me to leave. Though I didn't know it at the time, Sinclair and Acorn would soon be direct competitors; both trying to win the contract to design and produce a microcomputer for the BBC."

Cambridge Processor Unit

"I'd been thinking about starting my own computer business for a couple of years. Hermann and I had spoken about it at local watering holes and sometimes at the office in King's Parade. When Clive couldn't be persuaded to do more with computers, I decided to take the initiative by suggesting to Hermann that we start our own company. He agreed immediately. It was no way to launch a business – no business plan, no market, no capital, no idea what we were doing; but lots of enthusiasm!

"Hermann asked, 'How much does it cost?' I told him, '100 quid. Do you have 50 quid?' He thought for a moment. 'Yeah, I've got 50 quid. Let's start a company.' And we did. There wasn't any planning. There was just excitement about this technology that we felt was going to be big. We were young – I was in my early 30s – and ready to take a few risks.

"Shortly after Hermann completed his PhD, we incorporated Cambridge Processor Unit (in March 1978) as a consultancy business. We trusted each other and believed we could make it work. I stayed at Science of Cambridge because we couldn't afford to pay ourselves. At first, we built systems to order, aimed at the industrial market.

"Securing the first client was a novel experience – a fruit machine company in Wales. (In some countries they're called poker machines or slot machines.) ACE Coin Equipment was experiencing difficulties with their old mechanical machines as they coughed up the jackpot far too often. They wanted to develop an electronic version, where they could change the payback percentages. People at local pubs expected machines to pay back 90% of what was put in; any less, and no one would play these *one-armed bandits*. ACE usually installed their machines in transport cafés where people didn't spend hours playing them, so it didn't matter if they paid out less often.

"Clive had advertised that the MK14 could be used for all sorts of control tasks. 'It could even run a power station,' was one of his ridiculous claims. People from ACE called me at Science of Cambridge to ask if it could run a fruit machine. Seeing an opportunity for Cambridge Processor Unit, I asked Hermann to go down and speak with them. They had worked with some local electronics guys who had produced a machine that didn't work, but here was the expert from Cambridge…

"He agreed to look at the problem. He asked for a down payment, and they agreed. Chris Turner believes the total cost to ACE was about £20,000. That's probably about right. His memory for details is far better than mine. Besides, CPU's financial return shows it billed £10,000 in 1978, which is when a lot of the work was done. I expect we made the usual mistake of startups by not valuing our time. We were learning that a service business is a slow way to make a fortune. I doubt there was any profit. The fees might have covered our time, but we were working for wages and carrying most of the risk.

"To build a digital fruit machine, we needed people with skills we didn't have. Hermann turned to Chris Turner for help. He was willing to handle the engineering, but he needed help designing the processor cards and writing the firmware. The University Processor Group had about fifty members, but Steve Furber and Sophie Wilson stood out. Hermann told me that Chris, Steve and Sophie were the ones who could figure out our project – and they did. They would become the core of a great design and manufacturing business. Without them, Acorn never could have achieved what it did. Of course, this was long before they designed the ARM chip and changed the world. If you think that's a big claim, remember that the world's most successful product (the iPhone) is powered by the ARM chip. All you could do with a mobile phone before ARM was make a phone call. You couldn't even play Snake!

"We bootstrapped the company through this consultancy business. At first, we worked out of a back room at Science of Cambridge. It was a narrow house that had been turned into office space. ACE sent us one of their fruit machines to study. We couldn't get it up the narrow staircase, so it was parked in Pamela's and Hermann's living room for most of 1978. Meetings were held at King's Parade or at local cafés and pubs.

"We had a good relationship with ACE. They wanted an electronic fruit machine with a 'fun wheel' where they could dial up payback percentages. In their case, 'fun' was relative: When the machine was paying out, customers were having fun, but ACE wasn't making any money – and to them, that was no fun at all!

"It had two SC/MP microprocessors – one measured the position of the wheels, the other calculated the payback percentages; then it stopped in the right place. It was 'weighted' pseudo random, where the random number generator could be programmed to set the payback percentage.

"The software to control the two microprocessors was written by some of Hermann's contacts at the University Computer Laboratory. We didn't fully appreciate it at the time, but we had started to build a great team of computer scientists, mathematicians and engineers.

"Chris Turner was now working for Pye TVT but worked at home in the evenings and at weekends. Steve was doing his PhD in turbo machinery. We weren't allowed to pay him in cash because of the conditions of his research fellowship, so we paid him in kind – in electronic components. Just look at what he contributed to Acorn and the world – it must be one of the best deals we ever made. Sophie was a third-year undergraduate student, so she did the work in her spare time. Her degree results weren't as good as they should have been because studying became her spare time activity and working for us received most of her attention. But as history shows, it was time well spent.

"By August 1978, we needed our own office, so we leased space behind the electricity showroom off the Market Square at 4A Market Hill. It was much larger than what we needed at first and could be a little depressing, but we grew to like it. At first, Hermann worked there by himself, though Sophie, Chris, Steve and I were regular visitors. It wasn't long before we joined him full time."

Olivetti sort out Acorn

AT 2.30 in the morning of February 20, Hermann Hauser and Chris Curry, the two founders of Acorn, signed away their controlling interest to the Italian company Olivetti.

This combination – Acorn's strength in research and education with Olivetti's business and international markets – is seen as a potent combination.

Now the industry will look to Sinclair, which is not without problems, and Amstrad, as the two other large producers of micros in the UK for an indication of the state of the market.

Olivetti's stake will cost about £12 million, three-quarters of which will be used to allow for the reduction of value in stock caused by price-cutting, and losses in the USA and Germany. Acorn has announced that its total losses for the six months to January will be £11m.

This agreement will be put to Acorn shareholders at an extraordinary general meeting in the next month.

Olivetti, which is showing a series of TV adverts based on the Mr Men characters, will have a 49.3 per cent share in Acorn with the right to increase this to 50.1 per cent. Hauser and Curry's share will be cut from 87 to 36 per cent.

The whole structure of the company is being revised, and a new board appointed. A total of 120 people will be made redundant.

It appears Olivetti will support Acorn in going for education contracts in Europe, but - contrary to some reports

– Acorn is not pulling out of home computers. Development plans for the BBC system look safe and Electron production is continuing.

Elserino Piol of Olivetti stressed Acorn's 'outstanding technical expertise' and 'well established position' in the educational market. A joint statement by the companies added that Acorn will have access to Olivetti's international marketing to create 'a major drive in the world's education markets'.

There will be four new divisions within Acorn: education and training, scientific and industrial, business, and consumer.

However, the emphasis will be on the former to reduce Acorn's dependence on the 'volatile home computer market'.

Four present managers have been appointed heads of these sections. Hauser and Curry will become deputy chairmen, and the present acting chief executive Alex Reid will become chairman. Two other directors will be placed by Olivetti, and a managing director brought in.

This reprint from Acorn User, April 1985 reflects Christopher's understanding of Olivetti's commitment.

00000111

Christopher Curry – Acorn harvest

Christopher Curry had his sights set on an electronics career. By his mid-thirties, his focus had become less hands-on. In an interview for the October 1982 issue of *Practical Electronics* magazine, he showed his flare for marketing that influenced the success of Acorn Computers. He demonstrated a mature understanding of the market forces and the impending changes that loomed large on the horizon. Despite his understanding of what was to come, he couldn't prevent the inevitable. This concludes Christopher's story…

"The ACE project gave us some cashflow and confidence to think about making our own product – one we could *design once and make many*. We often talked about building a microcomputer kit for hobbyists. The MK14 had been a great success. Lots of people wanted them. Hermann and I thought about building something similar, also using the SC/MP microprocessor; but Sophie and Steve were unhappy with that choice and wanted to use the MOS Technology 6502. There was a lot of discussion about other microprocessors from Intel, Motorola, Zilog and National Semiconductor. We knew not to disagree with the self-appointed *President of the 6502 Fan Club* – 'After all,' Sophie argued, 'even Apple uses the 6502.'

"She designed the Acorn System 1 home computer, believing it could be made much better than the MK14; and it clearly was better. It had more RAM, more processing power, and was a lot more programmable. That's when I came to appreciate Sophie's talent. She produced the monitor program at astonishing speed – what we now might call an operating system. The entire monitor had to fit in just 512 bytes. She wrote every line of code on paper in machine code. There was one minor bug in the first version, but it worked well enough to be able to debug it. It was fully operational by the time we blew the second or third PROM.

"Needing a consumer-oriented brand for off-the-shelf product sales, we registered Acorn Computers in December 1978. Within a few months, the Acorn Microcomputer/System 1 was launched with a full-page ad in the March 1979 edition of *Wireless World*. We were on our way.

"It became the must-have home computer kit in Britain, because it had all the right features. Just like the Raspberry Pi and the Arduino today, people wanted to control lots of input and output pins so they could connect to whatever they wanted to control. It was ideal for that.

"It gave us cashflow and the confidence to recruit Chris Turner as our Chief Engineer – and first paid employee. Others soon followed. Sophie left the university and joined us full time. It was a bit harder with Steve. When he finished his PhD, he became Rolls-Royce Research Fellow at Emmanuel College; so, for quite a while he was a consultant. I joined full time in late 1979. By then 4A Market Hill was overflowing with people."

Funding

"Back in the 1970s and 1980s, the main source of seed funding was a bank loan. It was almost impossible to find anyone to invest in startups. We had no choice but to take out a loan. Hermann went to the National Westminster Bank opposite King's College. He told Mr Knight, the Manager, that we were starting a company, and needed working capital. After a short exchange, Mr Knight approved an unsecured loan of £10,000. No business plan to explain what we were doing, just a personal relationship with the banker.

"We managed to keep him happy because our account was usually in the black – at least once each month. Hermann went back to see him a few months later and said it's going very well, but we need some more money. Mr. Knight increased our line of credit to £50,000."

Atomic theory

"I had stayed at Science of Cambridge, even though I co-owned a company that was becoming a direct competitor. It came to a head after Clive left Sinclair Radionics, and we found ourselves sharing an office at King's Parade. Within a few months, I left and joined the growing payroll at Market Hill.

"I had been thinking about building something like the Apple II or the Tandy TRS-80. It wasn't long before I packed up and moved home, taking along Nicholas Toop, a talented hardware engineer, to set up a *skunk works* where we designed 'the next big thing' – the Acorn Atom.

"During 1979, Chris, Sophie, Steve and Nicholas had designed several System cards, so we had plenty of proven designs to draw on. We based our prototype Atom single board computer on those cards. Chris Turner did the engineering, and Allen Boothroyd designed the case. They got the Atom into production. It was our first big commercial success. By late 1980, we were shipping Atoms as fast as we could assemble kits and build complete units.

"The business had grown to a point where someone needed to focus on sales and marketing. The idea of making it look like a consumer product was compelling. Clive had taught me how to present products for the consumer market. I knew enough to do the advertising. Allen's injection moulded case looked fantastic. We had some very stylish photos taken of the mock-up. To finance it, we placed full-page ads in *Practical Electronics*. A few days after the magazine came out, the postman started delivering bags of mail with cheques enclosed, from people who were ordering and paying in advance for their Atoms. It was the first British home computer with a case and a full-travel keyboard, which users really liked. Clearly, we had an attractive product that people wanted to buy.

"We wasted no time getting it into production, though it wasn't all plain sailing. It had a US video controller that was not compatible with older UK televisions, but about 80% of TV sets could sync to the US format. As for the 20% that didn't work – well, we had lots of returns; but despite the video problem, it was a great success.

"The Atom wasn't the greatest computer of its time, but we managed to sell about 12,000 of them, and it helped to fund our rapid growth during 1980 and 1981. We were surprised that so many people would want it in kit form. The Atom had Sophie's BASIC, which made it much easier to program than machine code. It helped us get ready for the next big leap when the BBC came to call in February 1981.

"Our ability and confidence to think outside the box was growing too. Thanks to Andy Hopper, Paul Bond, Laurence Hardwick, Carl Dellar and Joe Dunn, we made one of our smartest moves by developing the Econet local area network. The education sector saw Acorn as an innovator."

A Proton in the works

"Steve and Sophie had been talking to Chris Turner, Andy, Hermann and me about what we should build next. We had different views about what to build: home, education, business, industry. Sophie suggested a dual processor machine; a multi-function computer that could run any microprocessor and operating system, a really radical idea at the time!

"The BBC television programme *Now the Chips Are Down* got people talking. Its success led to the *BBC Computer Literacy Project*, which aimed to educate the nation about what computing would bring. Their project team believed that the only way to do that was for viewers to get hands-on experience using their own computers.

"They'd been working with Newbury Laboratories, which had Clive Sinclair's failed design – the *NewBrain*. After Newbury failed to develop it, the BBC lost confidence in them. When I learnt they were unhappy with Newbury, I met with Executive Producer, John Radcliffe, and argued that it was in the BBC's best interest to open the field to other British computer companies – and the right thing to do.

"It took a couple of months for them to change direction, but they opened the field to six UK manufacturers, including Newbury Laboratories. We received a letter from John in early January 1981, followed by visits from the evaluation team who came with a list of features they wanted to see.

"Sinclair had been working on the ZX Spectrum. Clive made some silly claims and was quite unpleasant when speaking to the evaluation team. If you know the BBC at all, that's not the way to impress them. Three manufacturers, including Sinclair, were cut out on their first assessment.

"None of us had exactly what the BBC was looking for, but Steve, Sophie and Chris had been working on our next machine – the Proton. We thought the BBC specification was a bit ambitious for their budget, but Steve had a clear concept for our design, which was close to what they wanted. Once we realized that we could only win the tender by having something to show that was better than the Atom, we had no choice, we had to build a prototype Proton – in less than a week!

"Hermann doesn't remember why he waited until Sunday evening to call. Perhaps he needed to convince himself that it might be possible. We all agreed that a prototype would help convince the BBC to choose us.

"Sophie and Steve started on the Monday and spent the first two days doing the detailed design. Ram Banerjee agreed to help with the wire wrapping. Chris Turner used his powers of persuasion to get all the components together. It took Ram more than a day, non-stop, to wire wrap the whole thing. Hermann and I didn't get involved with the technical side; we made the tea, fed them, and gave them moral support.

"Because the circuit design was untested, it took all of Thursday and Friday morning to get it working. It took four days from start to finish; but everyone in the office contributed in their own way to help complete the Proton's design and assembly – a feat they can all be proud of.

"When the BBC arrived at 10 o'clock, they saw what had been achieved in just four days of development. They certainly didn't expect to see a working prototype from us when Newbury had nothing to show; so, they were impressed, and Acorn was selected to produce the BBC Micro.

"Though it was months before the contract was signed, we received confirmation from the BBC about a week after the demonstration. We threw ourselves into the project. At first, we focused on tying down the contract conditions. This would prove to be more of a challenge than we expected. There were several meetings with John Radcliffe, David Allen, David Kitson and Richard Russell from the BBC, John Coll from MUSE, and Mel Draper from the Department of Trade and Industry.

"It wasn't long before we knew we would need a lot more people. That would mean a big increase in overheads – nothing like we'd ever seen before. We hoped National Westminster Bank would continue to support us, which required another appointment with Mr Knight. 'Our business is going well,' Hermann said, 'but now we need a million.'

"Mr Knight nearly fell off his chair. He'd probably never been asked for a million pounds before, at least, not without any security; then their head office got involved. They looked at our company and couldn't believe we grew so fast, or that we needed that much cash.

"Negotiations were going nowhere but Matthew Bullock at Barclays Bank had developed a reputation around Cambridge – possibly the only banker in Britain who had been to Silicon Valley to study what was happening out there. He'd heard that Apple Computer was becoming successful, so he studied them and realized there was a big market for these home computers. He returned to the UK and saw the opportunity.

"Our only security was the BBC contract. In those days we could advertise the product, and people could order and prepay. We soon had 1.3 million pounds in an escrow account. We couldn't touch it until we delivered the goods, but the funds were there. The only risk to his million pounds would be if we couldn't produce the computers so he did his due diligence: He asked Cambridge Consultants, 'Does Acorn know what it's doing? Can they produce this computer?' They said, 'Yes they do, and yes they can.'

"Matthew phoned Hermann, 'Yep, we're going to lend you a million pounds.' – with no equity. So, we switched to Barclays Bank."

Let the games begin

"Before we could get far with the machine's design we had to agree on its specifications and other technical details. The BBC was flexible with the hardware, which wouldn't be difficult to sort out; but they had strong opinions about BBC BASIC – and so did Sophie. Her plans didn't involve slavishly following Microsoft. The BBC also wanted something that supported structured programming principles and was 'more user friendly'. Hermann got to chair that design group, which had Sophie and Richard Russell, John Coll, and a few other people from the BBC – excellent training for a career in the diplomatic corps.

"There were clashes between what John Coll and the BBC wanted with what we (well, Sophie) wanted. But they found ways to compromise; and BBC BASIC became a legend of a computer language in Britain that helped educate a whole generation of programmers. To his credit, Richard Russell kept BBC BASIC alive through his dedication and commitment to improving it and moving it to other platforms.

"The Proton was taking shape. It took a year of rapid development with all the stress and sleepless nights that entailed. It was to be called the *BBC Microcomputer*. The first machines shipped early in 1982.

"The BBC Micro sold well – perhaps too well. People had been hearing about plans for the television programme and government investment in education, so the computer received continuous media attention. Computers were expensive to produce back then. You could buy a new refrigerator and a washing machine for the price of a Model B. But despite the price, there was a lot of excitement and pent-up demand. People saved all year to have the money to pay for it.

"New government programmes introduced educational computing. The BBC Micro was adopted by about 80% of primary and secondary schools. Performance metrics for computer awareness in Britain placed it as a world leader for a generation of school leavers. Later, people in UK computer science departments lamented that youngsters stopped writing programs when the IBM PC replaced the BBC Micro in schools."

Coping with success

"When Acorn started shipping machines, the numbers looked promising compared to the 20,000 that the BBC had forecast. But then, the unthinkable: We received hundreds of thousands of orders while we were trying to sort out ULA problems, while we were blowing expensive EPROMs, and while John Coll was completing the *BBC Microcomputer System User Guide*. Channels to market didn't exist. Dealers hadn't seen a machine, nor had they received training to support, upgrade and repair them; so why should anyone be surprised about getting a lot of bad news at first?

"It was difficult to keep a sense of perspective as we were dealing with numbers that were mind-boggling: the number of orders for computers; the number of components to fulfil those orders; the number of invoices to process; and the number of people to recruit and teach how to perform their roles. Perhaps the number of customer enquiries (and complaints) was the hardest to deal with. Jim Merriman, John Horton, Peter O'Keefe, and others did a great job pulling it all together; but the people who faced the customers each day worked under constant pressure in circumstances that none of us could control. Slowly, each problem was sorted out, and the business settled down to become a well-managed operation.

"We never imagined how fast Acorn would grow, or the implications that would have. Turnover was about £10,000 in 1978 while Cambridge Processor Unit was getting off the ground. After Acorn started manufacturing computers, it all changed. Just selling Eurocard Systems, we billed £432,000 in 1979 – then several milestone were passed. We sold £1.4 million worth of Eurocard Systems and Atoms in 1980, followed by Eurocard Systems and Atoms worth £8.3 million in 1981. But it just kept climbing. We sold Eurocard Systems, Atoms and BBC Micros worth £42.3 million in 1982; and then Eurocard Systems and BBC Micros of more than £93 million in 1983. We were the fastest growing company in the UK, growing from zero to almost £100 million sales in fewer than five years – and we were very profitable. It was quite a ride but there were many times when we flew by the seat of our pants.

"By late 1982, we had been encouraged to float the company. Following a successful Initial Public Offering, Acorn Computer Group plc shares were valued at an opening price of 120p. The Group was born with a market capitalisation of £135 million. That took Hermann's original £50 investment to £64 million and mine to £51 million – on paper. Andy Hopper earned about £2.7 million.

"To put it another way: The value of the business grew more than one million times within those four years. Fast forward 40 years: With inflation the £135 million valuation would be worth £600 million or US$800 million. That was a very large company for the UK at that time.

"When the public company was formed, Acorn Computers Ltd remained the operational arm. Hermann and I ran it as Joint Managing Directors. He was also Chairman of Acorn Computer Group plc, while Andy Hopper and I served on the Board. David Johnson-Davies led Acornsoft, which produced great software. Cambridge Processor Unit changed direction when it became Acorn Leasing. Orbis Computers was a business that started life outside of Acorn but later joined us. It was set up by Hermann and Andy to promote Cambridge Ring technology.

"Vector Marketing Ltd had an unlikely start. It was a department within the Weetabix company that had a contract to distribute the BBC Micro. When they found themselves with some internal difficulties, Acorn acquired its staff and facilities, to continue looking after our distribution.

"We set up Acorn Computers International, with Bob Bayham taking control. He looked after Australia and New Zealand, and sold some machines into the Netherlands, South Africa, Hong Kong, Singapore, Scandinavia, Canada and the Middle East. India produced the SCL clone under licence. Acorn Computers GmbH was set up for the European market and Acorn Computers Corporation was formed in the USA.

"Later, Barson Computers Australasia Limited faced severe commercial pressures in Australia and New Zealand. Sam Wauchope negotiated with Julian Barson and his bankers to acquire the company."

Acorn marketing and advertising

"In the beginning, the BBC placed restrictions on our advertising. They eased up a bit later, but to be honest: there was so much demand for the first few years, it wasn't needed. Later, I wanted to raise Acorn's profile to help promote the Electron, particularly with teenagers and young adults. We sponsored a quirky promotion – the *Acorn Bus Racing Team.*

"The race series was run with Acorn-branded double-decker buses. It was a sideshow, but the public and the media loved it. The campaign injected fun into the brand and reinforced our image as innovative, but playful. It showed that while we had a reputation for research and development, we weren't afraid to be a bit absurd.

"We weren't serious about motorsport, though at Brands Hatch in 1984, we sponsored David Hunt, brother of Formula 1 champion James Hunt, in the Formula 3 event."

The Electron

"I prefer to talk about our successes, but we can also learn a lot from our setbacks. Hermann and a growing army of people dealt with day-to-day business activities. I felt I needed to focus on the future. The BBC Micro would not last forever, so I turned my attention to what should be next.

"Acorn needed a cut-down BBC Micro to compete in the fast-growing home and games computing market. It needed to run popular software for the BBC Micro, but only with those hardware features that consumers were prepared to pay for. The Electron had to be made as cheaply as possible to compete with Sinclair and new Asian products.

"We had our share of setbacks and disasters during the Electron's development. Perhaps we forgot how difficult it was to produce the BBC Micro's video ULA, or perhaps we wanted to believe it couldn't happen twice. But no, we used another Ferranti ULA for the Electron – the biggest available at the time – and we missed the 1983 Christmas rush.

"It was my decision to develop the Electron at the end of the 8-bit micro era – a huge mistake. Technical problems cost us a year, and the chance would never come again.

"As we couldn't deliver on time, the orders that WH Smith and others had placed with us were redirected to Sinclair. I admit it was a temper tantrum when I ran an advertisement in *Which?* consumer magazine, showing the failure rates of Sinclair in comparison to the BBC Micro. That ended in some much-publicised pushing and shoving with Clive in *Shades Wine Bar*. Ah well: There's no such thing as bad publicity!

"The Electron was about to become a huge problem as deliveries from plants in the UK and Asia were starting to arrive. Our warehouse was overflowing when we received an enquiry from what we were led to believe was a Russian company, wanting to buy 100,000 units. They wanted 1,000 immediately but would pay only £40 per unit. We were selling them at £70 to retail outlets and they cost £38 to build. We agreed to consider the deal on condition they were only sold in the Soviet Union.

"The 'Russians' came to see us at Cherry Hinton and placed a bank draft for half a million pounds on the table; and despite our protests, they left the draft with us. The next day they forced us to accept the deal, as one of our people at the meeting had failed to identify himself as our in-house attorney. To avoid an expensive legal battle, we accepted the deal. By then, we knew they preyed on financially stretched and over-stocked businesses like ours. They contacted our customers offering to sell them Electrons at £65. We ended up having to pay those fraudsters £120,000, just to get rid of them. Our attorney resigned."

Britain v USA

"One feature that made the BBC Micro so great was its input/output ports. You could connect just about any device to a Beeb and control it. That created problems for us in the USA. We made the same mistake a lot of manufacturers make when going into other countries: We ignored local requirements and the impact they could have on us.

"Hermann and I decided to go into the USA after we met a Sony executive on a stand at COMDEX in Las Vegas. We hired him to set up our Acorn subsidiary in Boston without doing enough preparation. Acorn had been very successful, with plenty of cash in the bank, which I think made us over-confident. We had been naïve, and we paid the price.

"To be fair to everyone in the Acorn US office, they never had a chance. I know they have been criticised, but they didn't have a product to sell. We lost valuable time trying to satisfy the FCC's emission rules. To pass their tests, we were forced to build the computer into a steel case, which delayed US and Canadian shipments for almost a year; and it made the machine heavier, more expensive, and less attractive to consumers. We did have some sales success, but nothing like what we had hoped to sell. One unexpected cost was having to engage direct salespeople who insisted on being paid an up-front retainer. That folly cost us millions.

"The writing was on the wall for 8-bit computers. We took them to their limits – twice the speed of other machines, like the Apple II and Commodore Pet; first to mix text and colour graphics, when most of the competition was only black and white; but processing power was always a limitation, and the 16-bit address bus could only address 64 kilobytes of memory.

"We were so busy we hadn't noticed that change was coming. Like Acorn with the BBC Master, Apple released the Apple IIGS to stay in the market while they worked on the Lisa, and then the Macintosh. Like us, they almost went broke. IBM saw it, and so did Commodore and Atari."

Project A is for ARM

"By the start of 1983, before the troubles with the Electron and the FCC emerged, we felt we had things under control. Steve and Sophie, with encouragement from Hermann, Andy and me, were thinking about what would be next. We knew that the education market wouldn't last forever; and while there weren't many small business computers in the UK at that time, it seemed like it might be our future.

"IBM's market share was growing, but we didn't anticipate the possibility of cheap PC clones being made by dozens of Asian manufacturers. It never occurred to anyone that MS-DOS would become a standard; until then, operating systems were tied to hardware, and vice versa. CP/M was the only exception that comes to mind. Unix had just been ported by the Wollongong Group – the first time it ran on anything but DEC equipment. IBM hadn't intended to allow Microsoft the freedom to sell MS-DOS to other manufacturers. They simply hadn't thought to restrict them. That oversight benefited PCs and compatibles. They've dominated business and home computing ever since.

"By the end of 1983, we had decided to develop our 32-bit RISC microprocessor. Sophie and Steve were authorised to recruit the design team – with no experience and few resources (though we did allocate £5 million to fund the project). Development of the first ARM chip started later that year. We were feeling confident about where we were heading with our technology when the business encountered rough times…

"Lazard was our banker when Acorn floated the company in April 1983. It sold its shares a year later when their price doubled. That appalling act of bad faith led to a massive crash in our share price. We should have sued but didn't. There were more challenges to follow."

The 1984 crash

"From the first day of 1984, the excitement of the IPO had been replaced by a nuclear winter for the computer industry. Atari was sold, Apple almost went bankrupt, the Electron had missed the Christmas rush, and demand for the BBC Micro collapsed. Acorn could no longer just rely on good luck. We were in a hole but kept digging, spending our reserves.

"At first, we struggled to make enough product to meet demand, so we could never get stock. We over-compensated by contracting more factories. As demand went through the floor, we were stuck with stock we couldn't sell and creditors we couldn't pay.

"The Group reported a loss of £11 million. Sales were dropping, but we'd placed orders with subcontractors that we had to honour. Meanwhile, we accepted orders from retailers that could be cancelled without any comeback.

"We kept prices high, then stock was devalued as unsaleable. Of course, we should have had proper contracts and sales agreements. We had a board of directors, and a very experienced senior management team. Why none of us asked the right questions is a mystery.

"With sales taking a nosedive, debt grew to a point where we urgently needed a solution. Olivetti was looking to invest in other businesses. They offered to purchase 49.3% of Acorn – which diluted Hermann's and my holdings – at a sale price of about £12 million. Alex Reid was appointed Chairman and Alex Uboldi later became the interim managing director.

"A week later, Acorn was forced to suspend share trading after its price dropped to 23p. Speculation about Acorn's future had been fuelled by Alex Reid's appointment as Chairman. Sacking Lazard as our financial adviser also unsettled the financial markets. Close Brothers replaced them; and then Cazenove resigned as our stockbroker."

"Even George Orwell couldn't dream up a more dramatic year for me. When it seemed that nothing else could go wrong, the Irish Republican Army (IRA) disrupted the UK Conservative Party annual convention.

"I was a member of the party and a great admirer of Margaret Thatcher, so I attended the Tories' 1984 Convention at the Grand Hotel, Brighton in October. It was very late, but I was still in the bar with other members.

"Fuelled by a little more to drink than was prudent, we were locked in deep conversation, debating the finer points of some aspect of government policy. At 2.45am, an IRA bomb, planted behind a bath in room 629 was detonated, taking out several rooms at the front of the building.

"We heard the explosion over the din of conversation in the bar. When it was clear that no fires had started, we all returned to our rooms and waited for further instructions from the security service. Later, I went to Brighton police station to make a formal statement. All I could tell them was: 'There was a loud bang, and I knocked over my drink.' We couldn't go back to the hotel because it was a crime scene, so I returned to London. A terrorist bombing is no laughing matter, but to be honest, I might have seen more if I'd watched the event on television. Even the news of the bombing didn't remove Acorn's continuing bad news from the front pages of the London daily papers.

"*The Financial Times* blamed our problems on an over-reliance on the BBC Micro and weak sales of the Electron. We called a press conference before Christmas 1984 to deflect rumours of poor sales and speculation about employee redundancies, being spread by our competitors.

"With no option, I was forced to admit that Acorn would soon launch a new version of the BBC Micro that had more features and would sell at a higher price. When it seemed like our public image couldn't get any worse, we cut the number of distributors and announced we would not be buying Torch Computers. In 1984 Acorn Group plc was valued at £216 million, but by the middle of February 1985, it had dropped to £31 million when we cut the price of our overstocked Electrons by £70. That hit dealers hard, making them anxious about shifting their stocks of Acorn machines, and equally nervous about taking on more stock.

"Acorn wasn't alone in this uncertainty. Commodore, Sinclair and other suppliers slashed their prices without warning. Established sales channels were under assault from all directions. By May 1985, High Street chains like Boots, Currys and Rumbelows were cutting prices, causing a feeding frenzy. Many small independent dealers went broke. I was desperately sad to see this because I had come to know many of them personally.

"The ARM project pressed on, isolated from the dramas going on around them. This was partly because the people in Project A kept to themselves to maintain secrecy. By April 1985 when the first chips were delivered, our world had been completely upended; but BBC Micros with ARM Second Processors were giving developers a taste of what was to come.

"When Olivetti took over, it scrutinised every aspect of the business and made changes, cancelling some product lines and business groups. Hermann and I had been looking for more than just a bailout. We had agreed to their offer because we believed they would sell BBC Micros through their worldwide network."

"Then, at a meeting in Milan, we learnt that Olivetti couldn't sell Acorn products through their sales channels, as they had agreed to do. Some of our senior managers (including me) were incensed by what appeared to be Olivetti's deception. It turned out there was a conflict with a deal they had with AT&T, which we didn't know about until it was too late.

"Olivetti was unwilling to do more to help Acorn, but with pressure from the Secretary of State for Trade and Industry, they agreed to acquire even more control for just £4,000,000. For them, it was the deal of the century. Hermann and I were both devastated.

"By the end of 1985, Alex Reid, Andy Hopper, Peter Wynn and I had resigned as directors.

"Independent of the changes at the top, there was a turnaround in Acorn's fortunes, with an increase of £10 million profit for the second half of 1986, compared to the same time in the previous year, due to internal realignments of the business and the launch of the BBC Master.

"Acorn gave up its plans for the USA and Europe. The BBC Master held on for a while, but it was just a matter of time. The world was changing. *Big Blue* had arrived and the only thing that could save Acorn was Sophie and Steve's ARM processor. That didn't happen without a lot of pain for everyone at Acorn.

"Hermann resigned to take up a position as Director of Olivetti Research in March 1987. Around the same time, the Acorn Archimedes was released – the first RISC-based home computer."

My life after Acorn

"A few Acorn engineers joined me to set up General Information Systems (GIS), in the old servants' quarters at my home – Croxton Park. We developed smartcard technology and other projects until late 1995.

"I'd started Redwood Publishing with Michael Potter and Christopher Ward in 1983. In the 1990s, it supplied data to network computers through Kingston Communication and Energis. Using a packet switched network was well ahead of its time and well before the Internet.

"The Acorn Communicator was a networked computer, based on BBC Micro hardware, inbuilt phone and modem. It could download data via Prestel services. We shrank the Communicator down to consumer size, calling it the Keyline terminal. Mail order companies, food retailers and other big retailers joined. Customers had a terminal at home and could ask questions like, 'Where can I buy my grandmother a present?' It would come up with a list from its internal database, and the customer could go online to talk to the supplier. It was a stepping stone toward the internet, but too localised to become viable. Before the world wide web started, it showed what was possible.

"My biggest regret is that I was unable to maintain the cost of that operation, as I tried to finance everything myself. Hermann was always much better at knowing how to use other people's money."

00001000

Andy Hopper

Though completely deserving of the title, 'Sir Andrew Hopper' sounds a bit posh for someone who likes to be called Andy; but that makes this British-Polish computer technologist and entrepreneur even more likeable. A remarkably active mind with the drive to push boundaries and make the most of every opportunity, that's Andy Hopper in a nutshell.

Acting as a conduit between Acorn and the extensive talent and resources at the University of Cambridge, he provided ideas for several innovations including early work on the Econet local area network and one the world's most successful chips – the ARM processor. As well as his many academic roles (including Head of the University Computer Laboratory), Andy became CEO of the Olivetti Research Laboratory in Cambridge in 1986. In this role, he established a series of successful spin-off companies that capitalised on the laboratory's advanced research.

In recognition of his contributions to British and global technology that created many billions of dollars in new industries, Andy was awarded a CBE and then knighted in the 2021 Queen's Birthday honours. He has been recognised with Fellowships of the Royal Society, the Royal Academy of Engineering, and the Institution of Engineering and Technology. Other awards include an Honorary Fellowship of Swansea University and an Honorary Degree from Queen's University Belfast, the Royal Society's *Bakerian Medal* and the Royal Academy of Engineering's *MacRobert Award*.

This is Andy's story…

"I arrived in the UK in 1964 at the age of 11 – unable to speak English, just Polish and some Russian. Within seven years I had become fluent and enrolled to study computing at Swansea University. David Aspinall had come from the University of Manchester Institute of Science and Technology (UMIST) the year before to take up his first chair of computer science. He had the imagination to put together an amazing course that was well ahead of its time.

University of Cambridge

"In 1974, Maurice Wilkes and Roger Needham interviewed me, then offered me a place in the University of Cambridge Computer Laboratory PhD programme. I studied communication networks with Maurice and contributed to the Cambridge Ring, a high-reliability high-speed local area network. David Wheeler was my Doctoral supervisor.

"Because I liked to do a bit of hardware design and construction, I was given space in both the PhD students' shared laboratory and with the technicians and engineers who had worked with the Capability Machine and with EDSAC 2. It was a little unusual for a PhD candidate, but I got to know the technical staff quite well. Later, I helped them with the Cambridge Ring. Maurice developed the conceptual ideas, David designed a lot of it, and Roger conceived the distributed system around it. I designed bits and pieces and helped the core of the engineering and technology team with its construction, while also doing my PhD. That involved designing a Very Large-Scale Integration (VLSI) chip interface for the Cambridge Ring and other types of local area networks.

"In the 1970s mini and microcomputers were coming onto the market. The challenge was linking them together without using telephone company wires and exchanges. The Cambridge Ring was one solution. Meanwhile, in the US, Xerox was developing the Ethernet, which worked differently. I made one or two contributions to the design of the Cambridge Ring and spent a lot of time on the VLSI chip. Government funding only paid for one person (me). Compare this to Ethernet, which had Intel, Xerox and DEC working on it. Looking back, it was madness to think we could succeed with such limited resources. Of course, as we all know: In the battle of the two technologies – Ethernet and Cambridge Ring – Ethernet was the clear winner. But there was a positive side: This work led to Acorn developing the ARM processor and other advanced technologies. It started with this project, but what came later was quite different."

My Parallel Careers

"After completing my PhD, I became a lecturer at Cambridge, where I saw my future climbing up the greasy academic pole. I've always worked in academia, but I was given an extraordinary amount of freedom to transfer technology to industry – both personally, and through my PhD students. For about seven or eight years in the late 1970s and early 1980s, I worked at the University while also helping Acorn. Building a parallel career in industry wasn't part of my career plan (because I didn't have a career plan), but that's what happened.

"It started when I met Hermann in 1978 at a disco at Darwin College. I like to say we were out chasing girls but that's a bit of a stretch. We went to the same pubs, talked about our work and what we should do next. Hermann said to me, 'Andy, you're doing practical work on Cambridge Ring; there's lots of interest from industry; why don't we start a company?' So, the idea of starting our first company, Orbis, came from him. He insisted we set it up 50:50. Hermann had the entrepreneurial drive, while I was still trying to make my Cambridge Ring chip work.

"Orbis needed someone full time. I was one of Mike Muller's lecturers and knew just how capable and trustworthy he was. Mike joined Orbis as its first employee, developing and building Cambridge Ring products.

"We had a couple of competitors. That was a good thing because the Ring couldn't succeed with only one company promoting it. Maurice sold the design with a non-exclusive licence. I think they cost £10,000. Orbis sold the board version while I worked on the VLSI chip design. Mike stayed through the transitions, moved to Acorn, and eventually became the Chief Technology Officer of Arm Holdings.

"We were surprised when Orbis got its first order so quickly – ICI Runcorn in Cheshire. The mark-up on the hardware was 1,700%, though we didn't put a price on our labour. Our client wanted to show ICI how a 'high speed' local area network at ten megabits per second could help their business. We worked our guts out constructing and delivering it and promised them a future technology pathway with the chip version.

"Hermann and Christopher set up Cambridge Processor Unit (CPU), a few months before Orbis. When they started Acorn, we shared premises at 4A Market Hill in the centre of Cambridge. We soon had CPU, Acorn and Orbis. Our small enterprise had become a conglomerate. Only our accountant benefited, so it seemed sensible to reduce the complexity. Hermann and Christopher were moving away from CPU's consultancy business, and it seemed sensible to merge Acorn and Orbis, so I sold my shares to them and became a minor shareholder and Director of Acorn. I was 'the third schmuck', as they say.

"There were only three Directors and three shareholders until it became a public company. I held the smallest share with 2% of the business. That's how it continued for the rest of my time with Acorn.

"Between 1983 and 1985, we offered Cambridge Ring interfaces with Acorn's Eurocard Systems 3, 4 and 5, and later for workstation projects. They were sold for highend or experimental systems, including some Acorn Cambridge Workstations running Unix. Sales were typically tied to bespoke installations in universities or research labs, though they were sometimes used to control critical infrastructure, as they were so secure. As Ethernet became the dominant local area network standard in the mid-1980s, Acorn stopped promoting Cambridge Ring and focused on computer manufacturing. Mike Muller moved to Acorn product marketing. Later, he designed the ARM I/O Controller and was one of the founders of Advanced RISC Machines Limited.

"The great thing about the University of Cambridge is the way staff can pursue activities in addition to their university responsibilities. We can accept external consulting assignments, have other sources of income, shareholdings, directorships, and so on. Most universities try to claim all intellectual output from their staff. That smothers entrepreneurial drive and industry collaboration. Cambridge has no restriction or barrier. It is one of few universities in the world to take such a liberal approach.

"Let me emphasise: This was a symbiotic relationship. For example, Acorn gave hundreds of machines to the computer science department. The transaction might be to give some of our time, or the use of software or other resources, and then Acorn might support our students. It could be something as simple as free pizzas for study, or whatever else we might need, so the benefits were two-way. It might seem like pizza currency, but it gave real value to us.

"Another example: My work as a PhD student designing the Cambridge Ring's first chip interface was useful for what followed. Building CAD software simulators was a simple but effective way to develop skills, knowledge and code that became available to Acorn for its own chip design. The idea of solving the BBC Micro's circuit board packing density problem using uncommitted logic arrays (ULA) came from the inventiveness of Steve Furber and Hermann, and from knowing what resources were available. Having that connection with the university helped improve their decision making.

"Acorn was designing chips for the BBC Micro and later the RISC processor, working collaboratively with the University. This was crucial for their future, and it enabled the University to learn about VLSI design. It gave us the confidence, toolkit – and people with the ability to do it."

Arm Holdings

"Success in business depends on being able to employ people who show good judgment, and who have the knowledge and confidence to say, I can do this. It's a judgment call with access to a tremendous amount of information. Mike Muller was one of the first ARM chip designers with Steve Furber and Sophie Wilson. He wasn't just the technological head of

Arm, he was also business strategist, with the right balance of optimism and aggression about products and technology; and look where it got them! Even after Arm was bought by SoftBank, he continued in that role. When the CEO changed, Mike never left, having been a key part of the company's success. He brought continuity to its development. In advanced technology, continuity is essential."

Olivetti Research Lab

"I was a director of Acorn until it floated, then I left and started to do other things. I came back as a Director of Acorn Computer Group plc up to the last few years before it was shut down. They ran out of cash, needed two rounds of funding, and became part of Olivetti. Out of a job, Hermann became their Vice President of Research, and one of the first things he decided was to set up a research activity in Cambridge. In 1986, he suggested, 'Hey Andy, how would you like to run a research centre doing what you've been doing for Acorn?' Of course, I said yes. I paid my university bosses the courtesy of asking permission though they seemed puzzled that I felt the need to ask. History shows that the University, Cambridge and Britain have done very well with this arrangement. When I was head of department for 14 years myself, I was able to help maintain the initiatives that we started at Acorn and at the Research Centre.

"The Olivetti Research Laboratory (ORL) was one hundred per cent owned by Olivetti but run independently with a separate board, with me as Managing Director. I maintained clear separation from this and my academic positions. Olivetti gave Hermann and me complete freedom to choose projects; and ORL enabled us to innovate using local talent, just as we had done when we started ARM development within Acorn.

"We recruited the same way we had done before – very quickly, drawing on top talent from the university. We looked at even faster networks, multimedia streaming – what became YouTube, webcams and that sort of thing. ORL quickly grew to about 50 or 60 people.

"Hermann's contribution was having a safe pair of hands at the helm. We decided that being under Olivetti's corporate structure wouldn't work; it needed to be a separate limited company that could do things the Cambridge way, with all the research being the property of that unit – symbiotic, quick recruitment, local processes – all of that. It became a separate company, of which I was the CEO. Hermann and Marco de Benedetti from Olivetti were on the Board.

"It permitted innovation in a way which, for all kinds of reasons, is not possible at a university. It succeeded because Olivetti understood very well how to leverage the local culture. They provided the framework that enabled it to happen. Herman and I proposed a way of working that would support spin-off companies based on the practical research and products ORL was producing. Olivetti took a 20% holding in spin-off companies. As a founder, I sought venture funding when needed."

Spin offs

"After a while, Olivetti said that what we were doing was too much for them to cope with. Hermann and I decided we had better start spinning out technologies. We first offered oven-ready opportunities to Olivetti business units. They accepted a few but most of the time, the teams raised venture funding and spun out the businesses themselves – the whole team, all the prototypes, and the rest of it. We open-sourced everything else because there was so much being produced that we wanted to keep alive and not just put in the bin. We weren't Olivetti salary staff, so we benefited a little from spin offs.

"Olivetti earned a few hundred million pounds from its investment in the Lab. Still, some of the Italians wanted to kill it off, but I asked for a chance to try and sell it. Olivetti agreed. It turned out that Oracle was interested. In 1997, it joined as a funding partner, and the lab was renamed the *Olivetti & Oracle Research Laboratory*. We were highly productive, not because we're brilliant, it's just that the structure made it that way.

"Advanced Telecommunication Modules Ltd (ATML), a Cambridge networking and communications firm, was the first to spin-off from ORL in 1997. It brought high-speed Asynchronous Transfer Mode (ATM) to the desktop but found the market slow to adopt the technology. ATML shifted to semiconductor design and licensing for broadband access technologies. This targeted Digital Subscriber Line (DSL) and broadband wireless markets, integrating communications software and chipsets. The sell-off was funded by a Californian Venture Capital firm.

"In 1998, when ATML re-branded as Virata Limited, it kept its strong R&D base in Cambridge but moved its headquarters to Silicon Valley. Virata Corporation floated on NASDAQ in 2000, raising about US$81 million. A year later, the company merged with Globespan in a US$1.3 billion deal to form GlobespanVirata, a major DSL chipset supplier. At its peak Virata's market cap reached US$5 billion. It was a remarkable thing: I was being paid a salary as managing director of the lab, yet I was spinning out, becoming a founder, and doing my bit for these companies.

"In 1999, AT&T was interested in starting its own lab in Europe. It bought ORL, renaming it *AT&T Laboratories Cambridge*.

"In 2000, Cambridge Broadband Networks was founded by ORL engineers, leveraging their point-to-multipoint wireless communications.

"For three years we were part of AT&T, but then it happened again! I received a call saying they were shutting us down. At that time, the US telecommunications industry was in turmoil. AT&T was looking for scalps. We were their only offshore unit, so we were an obvious target. As we held patents, there was a financial incentive. They had estimated what the patents would be worth if they went to market, while the book value of the operating lab wasn't worth that much. It seems that if they shut us down, they would receive a tax credit from the IRS worth tens of millions of dollars. (Who can understand the US Tax system?)

"This was at a time when the team was working on a project to develop a new broadband phone. Trying to prevent the shutdown, I approached MPs, the House of Lords, and even found two interested buyers: Intel and British Telecom. It felt like one of the worst moments of my life. Shutting the Lab was nuts. With very little effort there would have been considerable value there. It was simple logic: We were experienced. We had the talent and proven business methods. We had everything we needed to make it happen. It wasn't magic!

"AT&T was adamant that we had to be shut down, but there was a silver lining as they left so much value on the table. Because they made us all redundant, about six successful companies were born from their decision, of which I was a co-founder of three. One of AT&T's write-offs was sold to private equity for more than £100 million. Another was sold to a different private equity firm for £400 million. Yes, there was life in all this. We made good money and changed the world in our own way.

"While some projects were spun off into independent companies, others were transferred to the University of Cambridge Digital Technology Group within the Computer Laboratory. Successful projects included the Face Database for facial recognition, with other pioneering work in mobile systems, video conferencing, and location-aware computing.

"In 2002, the Virtual Network Computing project core team formed RealVNC Limited; then Ubisense Limited was formed to commercialise their realtime location system (RTLS) called Active Badge/Active Bat. In 2003, DisplayLink was established to build on its network display research of USB graphics/display technology. Then in the mid-2000s, Cognidox developed *Doxbox* document management for hightech firms – an internal tool spun off from ATML/Virata.

"Considering that America's greatest innovation experiment, Bell Labs, was an offshoot of AT&T, it's surprising that they took such a short-term view. American business culture versus Italian business culture – I know which one I have more respect for. To this day, we still have reunions from Olivetti Labs, even though it was shut down such a long time ago. People still turn up at the pub and celebrate the passion for projects in the lab; and the reason for it was Olivetti's strong business culture. It wasn't just permitted; it was encouraged by them."

The Cambridge Computer Lab Ring

"When I became Head of the Department of Computer Science and Technology in 2004, I decided not to participate in companies that were spun off from the department, to avoid any potential conflict of interest. We aimed to help as many companies as possible get started. To achieve this, we set up a business club called the Computer Lab Ring, a name suggested by Maurice Wilkes. The Ring is a networking group and a business club. There are mentors, people with money to invest, companies that have done well, and some not so well.

"To join the club, members must have been through the Computer Lab. They are given the opportunity to start their companies within the department with free office space and support. I'm proud that this has worked extraordinarily well, because it has helped hundreds of companies, with many of them still going. If you include Arm, which was a member, 18% of them have been sold for just over US$40 billion."

"It's Hermann's and my experience that we can't predict which ones will be successful, so it's no good trying to pick winners. The biggest missed opportunity was the iPhone. (Yes, we did the iPhone before Apple.) We designed almost the same device based on a small HP hand-held with a Wi-Fi connection and a pen screen, where we prototyped what looked and behaved the same as iPhone Apps. Triggered by tapping, we called them *web snacks*. I tried to convince AT&T to set up a development unit in Cambridge, but it fell on deaf ears, so that was a miss. The iPhone and its eco-system have become the world's most profitable product ever."

"I've retired from the University, which has a compulsory retirement age of 67. I'm an emeritus professor as they don't physically kick us out. I am happy to be at this stage of my career as it gives me time to pursue projects that interest me without having to always check my diary. I like helping to promote business innovation – companies like CommonAI, where I am co-founder and Chairman, and lowRISC of which I am also Chairman. They will keep me involved for the foreseeable future."

00001001

Sam Wauchope

Acorn Computers was forced to sell Olivetti a 49.3% share of the business in February 1985. A team from Olivetti spent six months analysing Acorn's financial position, which was in far worse shape than either party understood; and would need more than the initial £12 million bailout to survive. It was reluctant to tip in more cash after finding it had misread Acorn's position.

The Thatcher Government had been a staunch supporter of Acorn. Its collapse would reflect badly on the government, the BBC and Britain's emerging status in high tech industries. It happened that Olivetti was trying to secure a licence in the UK at that time. After some 'encouragement' from the Secretary of State for Trade and Industry, Norman Tebbit, Olivetti reluctantly agreed to increase its stake by 30.8% at a much lower valuation of £4 million – but also with other conditions. These included debt write-downs from Acorn's bankers and major creditors, which included the BBC.

These events led to Christopher departing and Hermann stepping aside from day-to-day management. A few of Acorn's directors and senior managers also left after some heated exchanges with Olivetti's senior management, where each blamed the other's perceived failure.

Sam Wauchope was a consultant with Arthur Anderson's London office when he accepted his assignment with Acorn, just as Christopher stood down as a director. He didn't expect this task to last more than a few months. How wrong he was! This is Sam's story...

"After attending a meeting with Brian Long, who had been appointed Managing Director earlier in the summer, I came to Cambridge in September 1985. It was the day of the first Board meeting after Olivetti agreed to provide additional funding. They intended to take a hands-off approach, though it nominated some new Directors and appointed Alessandro Uboldi de Capei as Chairman.

"On that day Christopher resigned. He was fuming when he left the very heated meeting. As we passed one another, he asked, 'Are you the new finance guy?' When I nodded, he replied, 'Well good luck, you'll need it.' With that, the man who conceived the idea of Acorn Computers, left to set up General Information Systems (GIS) and other ventures.

"Afraid Olivetti might kill the project or take it for themselves, it wasn't told about the ARM development until late in 1985. In truth, they probably wouldn't have been interested as the ARM chip had no place in their business, which focused exclusively on IBM PC compatibles.

"Brian Long's first task was to stabilize finances. It was a time when sales were slowing in Acorn's sweet spot, the education sector, though there was a buzz of excitement for people in product development, as the BBC Master had been launched and the first ARM chips had arrived.

"It was obvious to anyone with an ounce of objectivity that the company was overextended, with far too many speculative R&D projects. My initial role was to help Brian stabilize the financial position, starting by restructuring its leadership. It marked a shift from a business where its direction was based solely on the interests and ambitions of the founders, to management through disciplined corporate decision making.

"Brian came from Canadian tractor company Massey Ferguson. Though he had no background in high technology, he had held senior leadership roles in highly competitive environments that relied on engineering discipline and a strong focus on customer service. He moved quickly to dispel customer concerns about radical changes. 'Acorn's commitment to continuity sets us apart from our competitors and ensures loyalty from our customers. That is inherent in our success,' he said.

"It was early days for the personal computer industry – some of its most difficult times. Brian steered the business as it faced each challenge. In December 1985, he convinced me to move from my contract role to become Chief Financial Officer. Over the next two years we got sales moving, launched the Archimedes, and sustained ARM and some other development programs that were vital to Acorn's future – all while keeping finances under control.

"In December 1987 Brian left. Harvey Coleman, an Olivetti senior sales executive, was appointed Managing Director a month later. Harvey saw his role as maintaining the steady progress we had made, rather than initiating any major new programs.

"In April 1988, as part of a wider restructuring, I was appointed Sales & Marketing Director. It was an interesting change in a career that had been built on accounting and financial management, but I enjoyed the experience. The first Archimedes machines had been launched a year earlier and Acorn positioned itself away from its aging 8-bit machines.

"Two years later, Harvey moved to a new role. He distinguished himself after Acorn in senior roles including President of Olivetti Canada and later, President of Dell Canada. After his departure, I took over as Managing Director.

"It was a time for some big changes in the company. Most notably, six months after stepping into the role, the ARM project was spun off as a joint venture with Apple Computer and VLSI Technology to establish Advanced RISC Machines Limited. Apple liked the ARM technology but had reservations about relying on a small UK company – and a competitor at that. We had a team of very capable people who sold the joint venture idea to Apple and reeled the deal in.

"Recruited from European Silicon Structures (ES2), ARM Limited was led by Robin Saxby. Acorn retained a non-executive watchful eye on its protégé, which acted independently from the get-go and established itself worldwide without any significant input from Acorn.

"Over the next few years Acorn continued to sell into its core markets, although it failed to break out. It launched the Risc PC and its first laptop but also pursued other initiatives with a view to breaking into new and emerging markets. For example, Acorn explored handheld devices through a joint venture with Psion. It also established the first online 'Nursery' in Cambridge which developed broadband services. Banks, supermarkets, other commercial enterprises and schools took their first steps into what became the online world we know today.

"My regret is that we could not persuade Olivetti, still our principal shareholder, to allow us to raise the capital that would have taken these opportunities further. Olivetti remained locked in a narrow view of the PC market. They never saw the opportunities that taking Acorn products to their own markets worldwide might have meant to the success of both companies.

"By 1995, Acorn was facing another period of change and financial stress. It issued a warning about its first half-yearly profits in response to market pressures. I had come to Acorn as a consultant but found myself still there. I was weary of trying to persuade Olivetti what I felt should be our direction, and they decided I was no longer the man to pursue their narrower view of Acorn's future. I left in July 1995, and David Lee was appointed to the role. David was a chartered accountant who had spent 14 years with Olivetti, most recently as Director of Finance and Administration of Olivetti UK."

Note. In 1997, David stepped down and was replaced by Stan Boland who came from seven years in senior financial roles at International Computers Limited (ICL). The ill-fated Phoebe, the peak of Acorn's technical achievement, was created under his stewardship. He was also architect of the Element 14 spin-off. While he wasn't directly involved with the initial public offering of Arm Holdings, it provided the opportunity for Acorn to reach its final phase.

00001010

Chris Turner

Chris Turner had a boyhood interest in electronics, building homebrew radios, hi-fi amplifiers and even a few Sinclair electronics kits. This led to his first job as a technician in the Department of Physics at the University of Cambridge. He worked there by day and studied electronics engineering part-time at the Cambridge campus of what is now Anglia Ruskin University. One project was to design and build a high-speed image capture system for one of the research students: Hermann Hauser.

Chris moved to the Pye Group as an electronics design engineer in 1976, where he worked with advanced measuring instruments and microprocessors. While designing television studio equipment he worked with BBC Research. Exposure to television technology would prove useful in ways he could never have imagined at the time.

Disciplined and detailed, he kept extensive notes and recorded his initials on his work; so, 'CBT' appears regularly in old tech notes and schematics. Those records, his remarkable memory, and his attention to detail proved invaluable when researching this biography. Here is Chris's story…

"In 1977, Hermann was helping Christopher Curry develop Sinclair's first microcomputer, the MK14, which was based on a circuit design supplied by National Semiconductor. Hermann called and asked me to lay out the printed circuit board (PCB) artwork and write the user documentation. I started moonlighting for him in the evenings and on weekends in the spare room of my house. The extra income helped pay the mortgage. Hermann later recruited Steve Furber and Sophie Wilson.

"I agreed to help Hermann and Christopher when they decided to set up Cambridge Processor Unit in March 1978. Their first client, ACE Coin Equipment from Wales, wanted to put microprocessors into their gaming machines, which in those days were built with relays and electromechanical actuators. Steve and I started work on a prototype for an electronic fruit machine at Hermann's house in Grange Road on the 1978 May bank holiday weekend. I recall being in his front room, taking the back off an electromechanical ACE machine and staring inside. Not for the first time, Steve and I wondered what we got ourselves into.

"This was around the time telephone exchanges started to change from large, heavy, and unreliable relay switching to digital circuits. Old-style fruit machines were also filled with relays and actuators that looked like something from WWII parts salvage, so their makers wanted to go digital. Microprocessors that control solid-state switches (often power transistors, TRIACs or SCRs to drive motors, lights and other electrical loads) were starting to replace electromechanical equipment.

"We studied how the machine produced random combinations of symbols (usually pictures of different types of fruit) on rotating dials. Some combinations paid out prizes. Most didn't. We designed a system using two SC/MP microprocessor boards to do the same task. Steve designed the digital circuitry and I did the rest: input sensors for the switches, drivers for the lights, and some neat TRIAC drivers for the reels and motors. Sophie later added a radio frequency interference detector to stop people using piezoelectric lighters to cheat the machine by triggering a payout. Her solution was simple, effective, and well ahead of its time.

"Steve and I beavered away on the project in our spare time. By September 1978, the consulting business was in full swing. In December, Hermann and Christopher incorporated a second company – Acorn Computers. Hermann recruited people to write software for the fruit machine and to run simulations using the university's IBM 370 mainframe that predicted pay-out percentages. The project continued, but at a slower pace, and wasn't wrapped up completely until 1980.

"Sophie and Steve came up with ideas about what we should build next. Sophie wanted to build a competitor to the MK14 – but better. She designed our first consumer product, a microprocessor card with onboard RAM, ROM and some support chips, but no user input/output – the Acorn Microcomputer. Adding a second board with hexadecimal keyboard, display, input/output ports, and cassette interface made it the System 1. It was Sophie's PCB design and Steve's cassette interface. I drew schematics, wrote the hardware manual, and managed production.

"My part-time work had become more interesting than my day job. It consumed all of my spare time and offered more opportunities, so I left Pye in February 1979 and joined Acorn as its first full-time employee at 4A Market Hill. At the time, I was nervous about joining this startup, but excited about microprocessors and what I could achieve working with Hermann. He was probably nervous too, having an employee who expected to be paid every month, even if the business wasn't making money. I was the Chief Engineer, a post I held for several years. Eventually, the job grew too large for one person. I had to keep reinventing myself; as the company changed, my role changed with it.

"When the March issue of *Wireless World* magazine was published with its full-page ad for the System 1, we went out early to buy our own copies. Within days, we were overwhelmed each time the mail arrived. I still remember the bag of letters from people who sent cheques to pay for kits. Customers sent money with their orders, so cash flow was positive; and we shipped within 28 days (mostly) – a practice established by Sinclair and others. One day I received a call from a guy at a large company. He saw they were having a computer delivered from us and wanted to know how many men and what equipment he'd need to unload it. We retold that story for years afterward.

"It was the Cup Final weekend in May 1979. A year had passed since Steve, and I started work on the fruit machine. It was time to test what we'd built. I should have been home with my family, but I was down in South Wales where I spent the day feeding coins into a new all-digital ACE fruit machine and recording the pay-out results. I removed its cover so when I ran out of coins, I collected them from the back of the machine, then started again. I haven't put money into a gaming machine since.

"Many of the Eurocards we developed in 1979 and 1980 owed a lot to Steve's homebrew computer, but I did most of the detailed design, choosing components and finalising circuit diagrams. Of course, Sophie had a hand in them too, and she wrote most of the software. Pretty soon, there weren't enough hours in the day to get everything done, so PCB layouts were farmed out to GW Designs in Cambridge.

"I devised a product hierarchy and parts numbering system: 100 for assemblies and subassemblies, 200 for PCBs (I believe), 300 for software, and 400 for documentation. It continued with 500, 600, etc. for electronic and mechanical components. Each part number had an approved vendor list, which permitted the manufacturing sites to multi-source from what we identified as form, fit and functionally equivalent. If a site found an alternative, sometimes cheaper part, we'd have to approve it for use. From the beginning, we kept track of everything; and this early discipline paid off later as the company and product range grew.

"I've kept a lot of my notes. The pages filled with handwritten tables make me smile: They remind me what life was like before spreadsheets were invented.

"We produced more than a dozen different Eurocards – video, memory and I/O boards, disk controller, a backplane, keyboard, and a laboratory interface panel. There was a System 2, which came in a rack with a keyboard and video display interface. The System 3 and 4 had disk drives, and the System 5, with a faster microprocessor, was inspired by the BBC Microcomputer. Another development was our Prestel card which connected to a British Telecom service. With a Prestel adaptor, the computer could dial up the news, weather reports, chat boards, and a messaging service. It never took off but proved to be important when the BBC came to call because it used the same characters and block graphics as Teletext, which was so important to them.

"As we built more capable machines, other ideas took shape; one was networking. Andy Hopper was the inspiration for this, having worked on the Cambridge Ring at the university. The Econet local area network influenced Acorn's marketing for years to come. The Eurocard format was a smart move that allowed us to produce boards and systems that were sold for industrial control applications and Econet file servers, providing good cash flow. I believe I set the selling prices. Looking back, we probably sold them too cheaply, but we were keen to drive sales, so we had to keep things affordable.

"The Eurocard Systems, Sophie's BASIC interpreter, Steve's text editor, Nicholas Toop's EPROM blower, and the Disk Filing System helped to shape design and development. Being a bootstrapped company, Acorn could never afford expensive test equipment from Tektronix or Hewlett Packard, so we made our own development tools. Our in-circuit emulator (ICE) plugged into whatever product was being debugged, in place of its microprocessor, and could read from or write to any memory address or hardware register. It really came in handy when debugging prototype boards. Later, Steve and Sophie used a BBC Micro to develop the tools needed to design and test the ARM chip; a carryover from that time.

"There was a continuous parade of component sales reps as I ordered the parts we needed. Boards were assembled locally by RH Electronics in French's Mill – a cluster of small industrial units around a disused windmill. It grew into a successful business, making and selling third party add-ons. In the 1980s and 1990s, computer magazines were jam-packed with ads for RH, Watford Electronics, and other companies that were part of the Acorn ecosystem. The economic benefit to Britain was far greater than just the numbers that appeared in our financial reports.

"We made friends at the companies that sold us components. Some of them later joined Acorn and Arm. I think they sensed the excitement and potential. Sandy Dow joined us as our sales director from National Semiconductor. Before email, we had Telex. I still have a message from Motorola's sales rep, Robin Saxby, giving the timing data for some memory chips. Robin became the founding CEO of ARM Limited.

"Everyone pitched in to pack kits and process orders. New people helped to develop, test and generally get things done. Then as the business grew, we took on more technicians and other staff working in Market Hill and at French's Mill.

"I particularly remember Colin Priestley who was ever cheerful and could be relied on to get stuff built, packed and shipped under extreme pressure. He also helped me check through some of the first PCB layouts. I've heard he retired to Australia, living off grid near Ballarat in Victoria. It's just like him to live life his way.

"None of us were gifted salespeople, but we never seemed to frighten customers away. Like us, they were unsophisticated back then. We attended computer shows to promote our products. Our first show was in the Alexandra Palace with Hermann and David Johnson-Davies who brought some sophistication to our marketing efforts."

Acorn Systems and Atom

"Christopher had his own ideas about where Acorn should be heading – a home computer. With help from Nicholas Toop, the Atom was born. I worked with them on the production engineering. It was influenced by Eurocard System hardware and software but packed into a moulded keyboard housing. Allen Boothroyd, a local industrial designer, created the case for it, featuring a full-sized, full-travel keyboard. Christopher knew from his Sinclair days that a decent keyboard was a must, having heard so many complaints about the MK14's keypad.

"As the design came close to completion around April/May 1980, I sorted out parts lists, PCBs, and so on. I wasn't happy with the Atom as I expected there would be reliability issues with its voltage regulators overheating and the static RAM chips; but we pressed on and what we produced proved to be good enough. Later, I designed some of the Atom's accessories including the colour encoder and disk drive. A similar colour encoder design was used in the BBC Micro – it was based on a circuit I experimented with during my time at Pye TVT.

"The Atom was a great success that eclipsed Eurocard System sales numbers. They were sold in kit form or fully assembled and tested. It included Sophie's BASIC interpreter and an excellent user manual written by David Johnson-Davies. This was before he set up Acornsoft. Acorn sold more than 12,000 Atoms between 1980 and 1983 and could have sold more had it not been for the BBC Micro's release in 1982.

"I wrote the *Atom Technical Manual* which describes how to assemble the kit and how to connect peripherals. We offered a service to debug them when customers were unsuccessful assembling their computers. One of the Atoms that came back had all its components glued to the PCB because the customer didn't like to use a hot soldering iron!

"It was a great time to work at Acorn. The University Computer Lab was supportive of companies like ours. Maurice Wilkes, David Wheeler and Roger Needham wanted to see their research become commercially successful – preferably in the UK. Their doors were always open to us and many final year and research students came to work at Acorn, notably contributing to development of the BBC Micro and ARM.

"We weren't isolated from what was popular in Britain at the time: Rubik's cube was our favourite distraction. We listened to the Floyd and quoted *The Hitchhiker's Guide to the Galaxy*: 'Don't panic!' when we first saw the ZX80. Blake's 7, the BBC science-fiction series, used our System 1 as the control panel in its 'Slave' computer. Lunch at the *Copper Kettle* on King's Parade was a tradition, starting in the early days when Hermann, Sophie and I met up with Steve, who was working at the University Engineering Department, further along Trumpington Street. A lot of brainstorming and inspiration happened over those meals, and we sometimes enticed Steve back to the office to work on our projects in the afternoons."

BBC Microcomputer

"Many of my notes from December 1980 are from conversations with Steve and Sophie, with me trying to understand and capture their ideas for our next computer – codenamed the Proton. They had two priorities: It should be a dual processor system so we could run real-time I/O tasks on a proven 6502 micro, with a channel (later called the Tube) to a second processor. Also, the 6502 CPU and the video display circuitry should share access to high-speed dynamic random-access memory (DRAM). With DRAM, it's a case of use it or lose it. Simply by reading this volatile read/write memory, its contents are refreshed. So just having the video circuitry constantly accessing the DRAM also refreshes its contents. Using screen refresh to keep volatile data in the DRAM alive was Steve's idea, which he first used in his homebrew computer.

"By the time we returned from the Christmas break in early January 1981, I had drawn up their concepts while Steve wrote a description of what he wanted to build. At this point, the three of us were unaware of what the BBC was trying to do. If Hermann and Christopher knew, they hadn't mentioned it. There have been suggestions that we were influenced by the BBC's loose specifications because there were so many similarities. That isn't true. It's just that we shared the aspirations of many microcomputer users and developers at that time.

"Sophie developed specifications for a more advance Atom BASIC. It was John Coll, who christened it *Super BASIC*. What she conceived was close to what the BBC wanted to develop, though it would need Hermann's diplomatic touch to resolve differences of opinion between Sophie and the BBC's Richard Russell.

"David Allen and Dr Mel Draper from the BBC's evaluation team first called on us on the 27th of January. All we had to demonstrate were Eurocard Systems and the Atom. We had an outline of our next design – the Proton. We must have shown them our concept drawings as I understand they spoke about the Proton at a meeting in London a few days later. This was ten days before we formalised the design of the central components and built the prototype. It was a progression from the Atom, but a far more powerful machine with dual processor capability.

"DRAM ran faster than the Atom's static RAM and held a lot more data. The video was a complete re-design and our ambitious decision to use Uncommitted Logic Arrays (ULA) in the design would cause sleepless nights for some of us. Other improvements on our prototype Proton would take months of negotiations, wrangling and development.

"When we learnt that the BBC evaluation team was coming again on the 6th of February, we pulled out all stops. It took a heroic effort to get the prototype working in four days. Credit goes to Steve, Sophie and Ram Banerjee for their marathon effort, and to Allen Boothroyd who designed and mocked up the case – which apparently looked much better than other machines they had seen. But let's not forget Hermann, who kept us all sustained – in spirit, food and drink.

"I had volunteered my office for the week of all-night prototyping and debugging sessions. I couldn't join in as I had young children at home, but I got the components together, maintained the circuit diagrams and kept up with all the changes as the design evolved. On that first Friday in February John Radcliffe, Mel Draper and John Coll packed into my office at Market Hill with Steve and Sophie to see our prototype. The room was overflowing with people and equipment.

"Looking at the notes in my diary from that time, I'm reminded just how busy my job could be. It shows that I organised an Econet installation at a school on Hills Road, spoke to the college on Long Road, chased up GW Designs about PCB issues, sorted out ROM sizes and pinouts with NEC, and reviewed options for a new telephone system. (We were constantly outgrowing our phone lines.) I also arranged for a new TV aerial and licence, spoke to customers about System cards, met IC representatives, interviewed and introduced new starters to the company, and spoke to Bob Coates of the Microelectronics Education Programme about educational software development. (Bob joined Acorn in 1984 to run Education Marketing.) I can't say for sure, but I doubt I made it to the pub for lunch that week.

"Shortly after the BBC visit, I was in meetings with BBC engineers David Kitson and Richard Russell. We were mapping out the product specifications and scoping the work that needed to be done to get the BBC Micro into production. For the next few months, I maintained the master circuit diagram for the BBC Micro. We knew how big the board should be so I kept a close eye on the physical layout to ensure that the connectors mounted front and back would fit in Allen Boothroyd's case.

"As the design was fleshed out, we could see that we had too many components to fit onto a single circuit board. Steve was still gainfully employed at the university but would come in most days to talk through details about what would go where, while Sophie was dealing with BBC BASIC and other software issues. Steve and Sophie spent a lot of time thinking about the video modes, colour palette and pixel generator logic. A few parts of the design are mine, including the analogue interface, stitching in the Mullard SAA5050 Teletext character generator chip, some enhancements to the floppy and Econet interfaces, the detailed assignment of gates and flip flops, and the choice of connectors.

"Steve solved the component density problem using two uncommitted logic array (ULA) chips from Ferranti to replace the video and some clock circuits. The Serial ULA replaced the RS423/232 serial and cassette interfaces. Early versions of our schematics show the original TTL circuits for these functions with red pen lines drawn around what would fit inside the ULAs. Steve worked on them through the spring and summer. Andy Hopper, who had designed custom chips for Cambridge Ring hardware, enlisted the aid of Peter Robinson and Jeremy Dion at the University Computer Lab to create the artwork.

"Meanwhile, the circuit board went through early layout changes in readiness for the ULAs that arrived in the summer. Our draftsman, Bob Austin, did an enlarged layout using black adhesive tape and pads on mylar film, ready for photo reduction to make etch-resistant PCB production masks. Some PCB design software existed, but we didn't have it at the time, so the BBC Micro – with all its complexity – was designed using the old-fashioned tape-on-film method.

"I sometimes regret having named hardware address spaces Fred, Jim and Sheila, rather than calling them something more meaningful. I had three wires to get from one side of the circuit diagram to the other and no clear path to draw them, so I broke them and gave them names. I asked Sophie what I should call them, and she said, 'Oh I don't know – Fred and Jim.' (from *The Goon Show*). I added Sheila, which was simply the first name that came to me.

"I visited our plastics manufacturer in Manchester to see the first production cases. The hot plastic didn't fill the tool properly, so parts of the case moulding were deformed, which alarmed everyone at Acorn when they saw the sample. The problem was soon fixed, and complete cases arrived a few days later. There are examples of Atom and Beeb plastic cases in black and other colours before going to mass production, because 'first shots' (test samples) used whatever coloured plastic was in the machine at the time. These cases were grabbed by some of our engineers to use as their own computers and would sit proudly on their desks. I wish I'd kept a few because they're highly collectable now.

"Throughout development in 1981, the BBC kept a close eye on what we were doing and from my point of view, gave us excellent support without interfering too much. We held progress meetings over breakfast each Wednesday in the *Blue Boar Hotel* on Trinity Street with David Kitson, Richard Russell, and others from the BBC. I had a good relationship with David, who I think acted as a buffer for a lot of questions and issues, leaving us free from distractions.

"We fine-tuned the specifications with them, playing off cost and time against features and functionality. As I recall, the choice of red function keys was Richard Russell's idea. Sophie spent a lot of time fine tuning the specification for BBC BASIC with Richard. They achieved an excellent compromise that was largely compatible with Microsoft's BASIC, along with most of Sophie's enhancements such as structured programming procedures and functions. Paul Bond, a contractor from TopExpress, led development of the Machine Operating System (MOS). He did a brilliant job, making it modular so it could be written by independent programmers, contractors and Computer Lab students and researchers.

"I wish we'd made a machine capable of accepting plug-in boards, but we didn't have time to engineer it, and besides, most people wanted it to be cheap. When the first IBM PC was announced later that year, it had plenty of expansion but cost three times the price of a BBC Micro. Of course, there was plenty of expandability via the 1 MHz bus, the Tube, and so on; but considering all the extra boards people made to plug inside the Beeb, I wonder if we could have done more.

"Japanese memory chips manufacturers were in the ascendency. They had driven Moore's Law hard with their attention to cleanliness, quality and advanced lithography. The sales guys from NEC, Toshiba, Fujitsu and Hitachi were regular visitors to Market Hill.

"Hitachi's UK sales manager, Bob Daley, had spoken to me in December 1980 about their soon-to-be-released HM4816 DRAM. This was the breakthrough we needed to fit 32 kilobytes of memory into the BBC Micro's slimline case. Bob managed to get us some of the first samples for the prototype. He even drove up to Cambridge to hand deliver them to me. Timing was perfect. We couldn't have built the Proton six months earlier, as memory chips with those specifications didn't exist; but the semiconductor industry was progressing so fast and new chips were appearing all the time. Hitachi air-freighted chips to us as we needed them and they always supported us extremely well throughout the life of the Beeb.

"Shortly after moving into Fulbourn Road, Cherry Hinton, Hermann invited Bob Daley and the head of Hitachi in the UK to our new offices. Hermann, Steve, Sophie, other members of the senior management team and I met with them for a formal thankyou luncheon in our board room. We wanted to show our deep gratitude for their support. In typical Oriental style, everything was translated into Japanese. Our honoured guest's replies were translated back into English. At the end of the meeting, he stood up and said goodbye to everyone in perfect English.

"Some chips proved to be more challenging. Steve had chosen the Intel 8271 disk controller for his homebrew machine. It was used in Acorn's early products, with corresponding investment in Disk Filing System software. Intel couldn't supply enough for volume production, so we later switched to the Western Digital 1770.

"Ten years before the Internet and World Wide Web, we had Teletext – a system for viewing information on TV. It delivered news and weather updates, sports reports, recipes to accompany cooking programmes, and so on. We'd agreed with the BBC to add graphics support for Teletext (Mode 7), which wasn't in Steve's original Proton specification. BBC Research had also developed Telesoftware – a system for transmitting software programs over Teletext. They wanted us to design a Teletext Adapter so schools could download programs. It was high on their features list, so we held meetings with BBC Research engineers to determine how it would work. Mel Pullen kicked off the design work, but the concept was still evolving, so it took a couple of years to complete. Telesoftware was first transmitted in 1983.

"Prestel was a system from British Telecom using a modem to communicate over telephone lines; again, using Mode 7 graphics. It also published all sorts of data, and users could send messages and share information with each other. We already had Prestel working on the System 3, and we started the BBC Micro's Prestel adapter/modem project in May 1981, while the BBC Micro was still under development.

"With all the devices, greeting cards and cheap toys that now feature speech functions, it's hard to imagine that speech synthesis was only available in one consumer device 40 years ago: the *Speak & Spell* toy. The BBC Micro's speech option would enable access for the visually impaired and serve as a learning aid. Working with Texas Instruments, we produced a custom speech chip using the voice of BBC News anchor, Kenneth Kendall. I guess the work went smoothly because my notebook shows it was completed while I was away on summer holiday.

"1981 was a very busy year for me, pulling it all together, recruiting people and working out how to get 20,000 machines built (the initial contract commitment with the BBC). We believed that ICL in Kidsgrove would produce them, but for various reasons that didn't work out; then Christopher found a company called Cleartone in South Wales that was more flexible and better suited to handling contract manufacturing. I regularly travelled down to see them, transferring the design documents and working out how to get machines made and tested. Contract manufacturing is complex and takes a lot of effort. Acorn had multiple products and multiple manufacturing partners. Cleartone had multiple customers making multiple products. They purchased most components on our behalf and charged us for parts, handling, labour costs, machine time, overheads and profit. Their goal was to build exactly what I told them, even if what I told them was wrong. My challenge was to always give them the right information and ensure that's what was built.

"Planning the BBC Micro assembly process at Cleartone began at the end of August but production didn't get underway until October. There were two main sub-assembly suppliers: Astec and Wong's Electronics – both Hong Kong based. Astec made the TV modulators and later, the switched-mode power supplies. Wong's had made the Atom case and keyboard and would build the keyboard sub-assembly for the BBC Micro. Later, they both took on larger roles.

"No young engineer can be expected to walk off the street and start controlling a complex manufacturing project. I was fortunate to have worked at Pye Unicam and Pye TVT, where the drawing office was the central nervous system for the in-house manufacturing teams. Capturing and communicating the design, manufacturing, testing and maintenance for a new product needs libraries of drawings, specifications, instructions and other documents, along with a numbering scheme for parts and sub-assemblies, cross-referenced to component vendors' part numbers. For a relatively simple product like the BBC Microcomputer, this meant managing about 500 different components including chips, resistors, capacitors, connectors, fixings, and so on. For each of these, the organisation, development and configuration management of supporting documentation was almost as complex as designing the computer. Spare a thought for the people who build cars, aircraft and other complex systems, and you'll understand why documentation is king.

"So, the Beeb was designed, developed, and into production in just 12 months – delivering the first batch of factory-built machines in January 1982. In my view, this was a brilliant achievement by everyone involved, but we came in for criticism because not enough machines would be in the hands of customers in time for the first broadcasts of *The Computer Programme* in February. Looking at it objectively, the BBC had decided late in 1980 that their original choice of computer, the *NewBrain*, wasn't suitable and then searched for the newest and best technology to support the *Computer Literacy Project*. They did this while sticking as closely as they could to the planned broadcast date. Their visits and final demos in early 1981 showed the BBC engineers how our proposed hardware and software architecture, computational and graphical performance, expansion capabilities including Teletext, and all the other features would produce a machine that was head and shoulders above other contenders – ideal for teaching computing in many different applications. So, our yet-to-be-developed computer was chosen, and the BBC awarded the contract knowing we had an ambitious timescale before us. We got it done in 12 months and were able to scale production way beyond the BBC's initial expectation of just 20,000 units. As a result, the *BBC Computer Literacy Project* was based on a brand-new machine employing the very latest in technology and capabilities. That timeline should be hailed as a remarkable achievement.

"Cleartone started producing BBC Micros in November 1981, and by June 1982 had produced almost 23,000 machines – in line with the BBC's forecast. By June 1983 we had shipped almost 120,000 – an average of 10,000 a month. At our peak, we were making 5,000 a week worldwide. First production is known as a *Minimum Viable Product*. The fully populated Issue-1 board with one or two bugs still to be ironed out included a linear power supply that was subject to overheating. It had a softer plastic case that could be produced cheaper and easier, under-performing Ferranti ULAs with heat sinks, and a preliminary version of the Machine Operating System in EPROM. But Acorn stood by its product: Most machines with linear power supplies, early MOS ROMs and dodgy ULAs were upgraded, by Acorn or through its dealer network.

"Glyn Phillips helped me design and build custom test equipment: The Progressive Establishment Tester (PET) was a kind of in-circuit emulator that could sniff its way through the machine to help find faults. The Final Inspection Tester (FIT), connected to a finished computer, checked that it was all working before being packed for shipment. Final-test operators watched images on the screen to check its video modes and listened to test tones generated by the sound system – a cacophony of *Close Encounters of the Third Kind* rang out around the factory.

"I was particularly busy in South Wales at the beginning of 1982, dealing with a graveyard of dead Beebs that had failed final testing. Many of the faults were simple – such as a folded pin under a badly-inserted chip or a short circuit on the copper tracks of the PCB. These could take hours to find, but with help from Glynn, Robin Pain and others, Cleartone's engineers were soon up to speed and finding faults faster than we could.

"We also equipped our new dealer network so they could repair and upgrade the machines. For this, I wrote a service manual and delivered training courses – first in Cambridge, and then around the UK.

"With design authority for all aspects of the hardware, except the ULAs, I carried the circuit in my head and knew where every component was fitted. A continuing stream of engineering questions came my way – all needing a decision; but we had a capable team, so things got done. The main PCB quickly went from Issue-1 through to Issue-4 as we corrected errors, particularly around the printer port. Issue-4 ran for a while, then we moved through Issues -5 and -6, improving flow solder yield and getting a better fit onto our bed-of-nails test equipment, which detected PCB shorts and open circuits. I don't think Issue -5 or -6 went into production, so it wasn't until Issue -7 that we hit the final design.

"Factory acceptance tests kept me away from home and the office in 1982. I often travelled with a BBC Micro in my suitcase to check first samples, and to give authority to start production. There were regular visits to Hong Kong. At Wong's Electronics, we ramped up an extra production line, along with an updated version of the case that was made from better quality ABS plastic. I remember sitting with a group of Wong's engineers measuring a first shot of the new case with my verniers and pointing out errors. Their response was, 'OK, no problem, we fix.' and they did, by the next day!

"Astec produced a more efficient switched mode power supply to replace the overworked analogue design (which the BBC had insisted on, against our advice). I worked with VLSI Technology in California to replace the Ferranti ULAs with CMOS gate arrays; where, without warning, I had to invent a way to invert the Mode 7 cursor function.

"Long-haul travel in those days was quite different – no mobile phones, fax machines or laptops. Once on site, with only a crackling telephone or telex to stay in touch with Acorn, there was no room for errors and few opportunities to discuss issues. When I visit California now, I see how much it has changed. When I left Highway 101 in the early 1980s, heading for VLSI Technology's new facility on McKay Drive in San Jose, the road passed run-down bungalows and fruit orchards. Nowadays the trip passes smart buildings housing hi-tech companies and mansions."

"British computer development was in the doldrums during the early 1980s; and may be why the BBC Microcomputer was such a hot topic. Speculation extended to professional and academic circles. As an active member of the Cambridge branch of the Institution of Electrical Engineers (IEE, now the Institution of Engineering and Technology), I was invited to present an evening lecture to introduce the BBC Micro. Steve, Sophie and I went along in November 1981 with a pre-production machine hooked up to a projector. About five hundred people packed into the University's Engineering Department lecture theatre. We weren't trained presenters, but they must have enjoyed it as there were follow-up invitations to present at their London branch, and later at their Savoy Place headquarters. That lecture was oversubscribed, so we appeared twice more. After that, we toured IEE branches across the country. In Cardiff, I recall a beautiful theatre with carved red dragons across the roof. Senior management from a company called AB Electronics filled the front row. They bought Cleartone the following week.

"We enjoyed presenting those lectures. They created interest among industry professionals and widened the BBC Micro's appeal with people in science and industry, for education and personal use. I remember a dinner with the IEE people in London after the January lecture when an enthusiastic chap asked Steve what we would do next. Steve contemplated for a moment and then replied, 'Have pudding, I hope.'

"After two years of fighting for space to work, we moved to a converted water treatment plant on Fulbourn Road, Cherry Hinton at the end of 1981. Finally, everyone from Market Square and Bridge Street could be in one place. Development and manufacturing activities increased. Jim Merriman recruited the two Davids: Hughes and Ireland, as production engineers. Phil Smith managed quality assurance and Barbara Cole ran purchasing. Once they were up to speed, my life became a little less frenetic and I was able to focus on research and development.

"The headcount continued to grow so a cluster of portacabins appeared. Acorn became a public company in October 1983, about the same time as a new R&D building was constructed. It immediately filled with engineers. Christened the 'Silver Building', it was repainted yellow in 2020 as one of Arm's buildings. It was only meant to last 20 years but stayed for 40 – until they replaced it with a multistorey car park.

"Acorn's 1984 Annual Report stated that the total headcount had grown to 450 worldwide, with 300 located in Cambridge and over a third in R&D. This number included Acornsoft and the Acorn Research Centre in Palo Alto. The years 1983 and 1984 were challenging. There was a lot going on and as Chief Engineer, I was managing a large part of our R&D division with support from Alan Fournier, Dave Dicken and others in my team. I interviewed and recruited many of our R&D engineering staff, which would include some who would later become founders of ARM."

The US Beeb and Domesday Projects

"My greatest technical challenge was developing the US version of the BBC Model B, where we failed to meet the FCC radiation and UL safety compliance standards. The US regulations were difficult to achieve as the BBC Micro wasn't designed to meet their rules – which most countries now adopt. Radio interference measurements had to be made with a one-metre cable connected to all connectors. The Beeb was too expandable for its own good and radiated like a beacon. Richard Hughes added metal shielding which fixed the problem but increased the cost.

"With cables strung up in the air and a complex antenna system a few meters away, test measurements were performed beside the Silver Building to get away from other interference. Nearby residents complained to Council, so I had to reassure them we were not exposing anyone to harmful radiation. We did finally achieve compliance; but too late for the US Beeb to become a commercial success.

"Pushing boundaries created challenges like the Electron ULA, which delayed its introduction, losing market opportunities, then creating a cash-flow crisis. Olivetti took control of Acorn in 1985. Before long, we downsized and restructured under a new CEO, Brian Long. I became Operations Director in the Custom Systems Division run by John Horton, who had joined us in 1982 as Technical Director. We designed the BT CHAIN terminal, a rack-mounted version of the Beeb for Reuters, and the Acorn Communicator. We also ran Acorn's interactive video disc business based in Maidenhead.

"As a change of pace, I took on the BBC *Domesday* video disc project and delivered the interactive video system on time in November 1986. I also became the eyes and ears for compact disc technology. CD-ROMs were regarded as the 'next big thing' in 1987. I sponsored development of the first CD-ROM interface and file system for our new Archimedes computers. But with storm clouds on the horizon, another downsizing and reorganisation, I left Acorn early in 1988 to join the new Olivetti Research Lab with Andy Hopper. This marked the beginning of the next stage of what has been a rewarding career in electronics engineering."

Reflections

"The MOS ROM in the BBC Micro lists the names of people who contributed to its development. I've mentioned a few of them, but there are many more friends and colleagues – too numerous to mention – who have earned my gratitude. It's important to remember that when we received a Queen's award for technological achievement in 1984, every individual, regardless of their role, who had been part of Acorn Computers, could justly claim to have contributed to its success.

"Our industrial designer, Allen Boothroyd, deserves special mention. Allen could think in three dimensions – what a remarkable guy with some equally remarkable skills. Sadly, Allen passed away in 2020. What he brought to Acorn contributed to our success. For that, we are grateful.

"Reflecting on our work in the 1980s: we designed the Beeb in a golden age. Consider EDSAC at the University of Cambridge in 1949. It contained 3,400 valves and consumed 11,000 watts of power. A large team worked for three years to deliver that machine. Today, we have chips developed over several years by huge teams, reliant on complex silicon tools and sophisticated fabrication technologies that cost tens of millions of pounds. These didn't exist when Steve, Sophie and I – in our spare time – created systems using off-the-shelf chips that excited people's imaginations. So, in terms of the technical impact that individuals or small teams could achieve, I believe we were there in a golden age. It's reflected in how much people could achieve with their BBC Micros – immediate and impactful results, without needing large teams of people writing massive programs. It spawned a generation of embedded systems experts who still work in electronics and software today, designing and programming machines to perform tasks in real-time. That is our legacy."

00001011

Steve Furber

Professor Emeritus Stephen Furber CBE FRS FREng DFBCS FIET CITP CEng was appointed to the ICL Chair of Computer Engineering at the University of Manchester in 1990. After a Rolls-Royce Research Fellowship at Emmanuel College, Cambridge, he joined Acorn and led development of the BBC Microcomputer and the Acorn RISC Machine (ARM) microprocessor with Sophie Wilson.

Professor Furber is a Fellow of the IEEE, the Royal Society, the Royal Academy of Engineering, the British Computer Society, and the Institution of Engineering and Technology. His awards include a Royal Academy of Engineering Silver Medal, a Royal Society Wolfson Research Merit Award, and the IET Faraday Medal. He was awarded a Commander of the Order of the British Empire and was a 2010 Millennium Technology Prize laureate, awarded by the Technology Academy of Finland. He is a Computer History Museum (Mountain View, CA) Fellow Award honouree and has honorary doctorates from the University of Edinburgh, Anglia Ruskin University, and Queen's University, Belfast. This is Steve's story...

"I went to Manchester Grammar at the age of ten, then received a scholarship to enter St John's College, Cambridge. After completing my Part III Maths, I changed to the engineering department to complete my PhD in aerodynamics, looking at ways to improve the efficiency of jet engines. My thesis was submitted in 1979 and awarded the following year. I was at Emmanuel College, on a Rolls Royce Research Fellowship until 1981.

"People were starting to play with microprocessors and electronics. At a local pub I heard about the Cambridge University Processor Group (CUPG). This interested me because I thought it would be fun to build a flight simulator. I needed a computer to do that, so I went along to a meeting to see what was going on. It was founded by people who enjoyed building computers.

"The 'real men' used discrete logic to build theirs, while wimps like me used microprocessors. We ordered chips by mail order from California, paid for using our new-fangled credit cards. We'd build a basic computer, then we'd go to each other's houses to show them off, and to get help with debugging. That's where I first met Sophie Wilson. She enjoyed finding bugs in my computer.

"Hermann Hauser wasn't on the scene at that time. He was a postdoc student in Physics at the University's Cavendish Laboratory. He had started talking to Christopher Curry about forming a consulting business, because microprocessors were clearly going to be important. Christopher was working for Sinclair at the time, but he was frustrated with Clive's reluctance to let him develop a microcomputer, so he and Hermann cooked up the idea of starting Cambridge Processor Unit (CPU). As the idea took shape, they thought that the CUPG would be a good place to look for people who could do the technical work. Hermann found me through the group, though he wasn't an active member. This was the first time he demonstrated his ability to pull resources together.

"Hermann and Christopher were beginning to put their ideas together while looking for people who could help. My first project was to design a twin-processor controller for ACE Coin Equipment's fruit machines. I think the first meeting, where they talked to us was in the Fort St George pub on Midsummer Common early in 1978. Chris was working for Pye. Sophie and I were at university.

"Under the terms of my research fellowship, I couldn't do extra paid employment but what they were doing interested me. I designed bits and pieces for them, and they gave me more bits and pieces to play with in return. There was synergy between my work for Acorn and my University day job as I used the computers in my aerodynamics experiments.

"When Science of Cambridge produced a DIY kit called the MK14, Chris Turner had done all the work. Sophie Wilson took one look at it and said those words that we would hear more than once, 'Err, I can do better than that.' She was designing what she called the *HAWK*. After seeing her ideas, Hermann and Christopher decided they would prefer to sell computers, rather than run a consultancy business. Sophie's design became the Acorn System 1, which was marketed as a kit, and sold in quite reasonable numbers. That changed the whole direction of the business.

"When Chris Turner joined full time, he brought professional engineering and manufacturing discipline where there was none. When Sophie joined, she took on programming and helping wherever she was needed. I didn't join the company until late in 1981, so I didn't get involved in day-to-day activities before the BBC came along.

"One of the philosophical ideas of this period was that the architecture of machines should be as open as possible, so you could plug whatever you liked into it. That was one of the things that expanded this market so quickly: Others could build peripherals and add-on devices for them. It later influenced our thinking about the design of the BBC Micro.

"The BBC had a reputation as a huge, monolithic and slow-moving organisation; but influential people at the BBC and in the Department of Industry could see that computers would change the world and wanted to help the wider community get ready for the future. As I later learned, they had been looking around for someone to build them a microcomputer to accompany a series of TV programmes for almost a year. They'd been working with Newbury Laboratories, trying to develop an affordable and capable computer but were frustrated with the lack of progress of Newbury's *NewBrain* machine. In December 1980, they decided to consider other types of computers.

"We didn't know about any of this when we started talking about our next design. The Acorn Atom had sold well, showing there was a market for a user-friendly home computer. Some insiders wanted a 'better Atom', some wanted a machine for business, and others wanted a machine for laboratories and other university applications. Sophie came up with the idea of having a basic machine to handle input/output – something that would suit home and school computing – with a communication channel to a second computer that could do the heavy lifting. The channel would become 'the Tube'. There were a few technical reasons why we couldn't use a bus. It occurred to me while traveling on the London Underground: If it's not a bus (or a taxi), it must be the Tube. That's how I came up with the name for the connection between the I/O processor (which we had codenamed the Proton) and the second processor (which we imagined would contain a 16- or 32-bit micro).

"Christopher was able to get the BBC's attention. What we didn't know at the time was that they were desperately looking for a UK company to show they could build what was needed, when they needed it, at a price the public could afford; and that the successful company would not do anything to embarrass the BBC. We knew that the Acorn Atom wouldn't be acceptable, so we needed to build on what we had and to come up with something that was a lot better – like the Proton. Without making any commitment, and not knowing what we had to show them, they agreed to come to Market Hill on the following Friday.

"We didn't have a design, let alone something to start building. We had a sketch – little more than a block diagram for the Proton. We didn't even have a clear idea how our Tube would work. Like a lot of the machine's design, it would not be without its challenges. But if there is a single feature that made the BBC Micro unique, that was it.

"It was the first Sunday in February when Hermann called to play Sophie and me off against each other saying, 'The BBC is coming on Friday. Can we show them a prototype?' Despite our protests, he persuaded us to build one in four days. Time was tight, but with luck, it might just be possible.

"I spent two days sketching a preliminary circuit diagram and checking the details with Sophie. By the time Ram Banerjee arrived on the Tuesday evening to do the wire wrap, we thought it was as good as we could get it on paper. As it turned out, a few errors were spotted as the prototype was assembled in wire wrap.

"Chris Turner had worked his magic to source the components we needed. Hitachi came through with the only available memory chips in the country that were fast enough to meet our specifications. Ram worked for almost a day and a half straight without a break. Some of us took turns assisting him by calling out which pairs of pins needed to be connected and the colour of the wire to be used – based on his system of colour coding. Hermann and others kept up a steady supply of food and drink. Early Thursday morning, Ram left to catch up on some well-earned rest. It was now up to the rest of us to get it working.

"We worked all day and late into the night, slowly finding and removing bugs. There weren't many wiring errors. Most bugs were due to a few errors and omissions from the design. There had been no time to thoroughly bench check the logic before we started prototyping, so we were doing it the hard way: on the live circuit. Slowly and methodically, we checked every part of the circuit using our homebrew 6502 In-circuit Emulator (ICE). We checked the power supply voltages and clock, control lines, and the data and address buses, the reset and interrupt lines, and signal timing. Everything seemed to be operational – the signs of life in any computer. We monitored the address bus as it sequenced through binary addresses. The buffers – chips that allow several devices to share the same bus – all checked out. We confirmed that the memory was saving and loading data.

"I don't recall how much time I spent there but I still had my day-job at the University and a wife at home who liked to see me occasionally, so I was relying on others to keep the work moving. I was back working on it on Thursday evening. We worked through most of the night debugging the protype but according to the ICE, there still seemed to be a problem. In the early hours of Friday morning, Hermann applied Occam's Razor

– the problem-solving principle that recommends searching for the simplest explanation requiring the fewest assumptions. As he put it: 'If everything seems OK but it doesn't pass the tests, maybe the ICE is at fault. Let's remove the ICE and test it with the microprocessor installed.'

"It worked!

"We all collapsed in a heap – wherever we could find some floorspace. When Sophie arrived early on Friday morning, we were all fast asleep. She installed the demonstration software with just minutes to spare, then the one and only Acorn Proton ever built purred into life. We were ready to show what it could do.

"Hermann had recruited Allen Boothroyd to design a mock-up of the case. When the people from the BBC arrived around 10:00AM that morning, we had the prototype running code, and we had a case… 'This is how it will work; and this is what it'll look like.' Clearly, the fact we had done that in a week persuaded them to choose Acorn.

"What followed was a very busy year for everyone in the development team. We needed to add a few people to get everything done in time but overall, our small team worked well together. My focus was getting the uncommitted logic arrays (ULA) designed, and the problems sorted. Having Peter Robinson and Jeremy Dion at the University Computer Lab was a stroke of luck. We could never have done it without them. We did have some problems with the Ferranti ULAs, both with the video processor in the BBC Micro, and later with the Electron ULA. In both cases we were working too close to the limits of the technology.

"Getting a complex job donc with a lean organisation is easier than getting it done with an excess number of people. Of course, it would be different if we were building a super tanker. If the size of a team is smaller, communication is easier to manage, with fewer misunderstandings. I'm not against big teams; I've just never worked with one. In my experience, a few talented and committed people can achieve some remarkable things. That's how it was with the BBC Micro in 1981, which had perhaps 30 people. We all understood what was at stake and knew we could rely on each other. That made us so much stronger; and later, when Acorn grew to several hundred people, the ARM design team had ten. Because I've been at a university with limited research grants, we still had to work out how to do things with quite small teams. But times have changed: designing a processor today, needs at least 100. I have no experience working with that many people. I wouldn't know where to start.

"Where Acorn needed more staff was supporting its customers. The BBC expected to sell 20,000 machines on the back of their TV programmes, but according to the website *bbcmicro.computer*, about one million were shipped. Nobody anticipated the public's appetite for computing. The BBC quite rightly felt it their duty to support this initiative, but I don't believe they had done any serious market research.

"There weren't any alarm bells. The Atom had sold about 12,000, overall. The number that the BBC predicted seemed quite reasonable to all of us at the time. Chris Turner, Sophie and I began to sense how big this could be when we conducted a series of over-subscribed seminars around the country with the IEE. Computers were becoming consumer products, with a lot more finding their way into homes. I was puzzled by this. Though I had a computer at home, I couldn't see why most people would want one – but obviously, I was wrong. At first, they were just a curiosity, but people could see that their children would need to be computer literate, so BBC Micros became more common in schools and then in the home. People just wanted to familiarise themselves and see what these things would do. Very rapidly of course, other factors came into play, and one of the big ones was computer games.

"When we started dealing with the BBC, it became clear that the government would give significant backing to the *Computer Literacy Project*, so we were going to have our machines in schools in quite a big way. Then we were going into homes as it gathered its own momentum. In no sense were we pushing it along. Yes, Acorn did some advertising, but compared to the number of machines being sold, the marketing budget was very small.

"We were standing back, watching, and trying to anticipate what the next move would be, so we could be ready. We made commitments to the BBC to develop ancillary devices. I was responsible for the 6502 second processor, which could plug into a Beeb and make it at least two or three times faster, and far more powerful. There would be other second processors for commercial applications, and the Teletext and Prestel boxes to download telesoftware over the airwaves or telephone lines. The BBC Micro had a lot of interfaces that spawned applications for classrooms, laboratories and process control. Some were from third parties, but there was a lot of activity from Acorn, as well.

"The organisation was run by Christopher and Hermann, and they scaled to that job quite well. They employed experienced managers to support them, so it wasn't just members of CUPG running the company. The people brought in from outside Acorn had manufacturing, financial and other management experience to run the company as a serious going concern.

"There is no argument that Acorn was caught napping when it attempted to launch into the US market. It wouldn't happen today because there would be a task force planning every part of the operation. Threat assessments, marketing analysis, scheduling and project plans. Acorn had very little of that. The people involved believed they had everything under control but clearly, they didn't. Back then, UK companies took too much on trust and probably made too many assumptions. It cost a lot to discover that FCC regulations in North America had much tighter rules on radio emissions than in Europe. They were a real impediment to building the BBC Micro with unterminated connectors. Radio frequency interference was not yet a problem in Europe, but it was just a matter of time. Any suggestion the policy was designed to block foreign companies is nonsense. It was a sensible rule, because radio emissions from electronic equipment were becoming a problem that needed to be controlled; and those controls are now international. We never did build a fully compliant BBC Micro at an affordable price, which is why we came home with our tails between our legs.

"The next step up from the 8-bit machines was to build a 16-bit microcomputer. Like everyone in the industry, we were subject to Moore's Law. With functionality growing so fast, going from eight to sixteen bits seemed a natural progression. We looked at the 16-bit micros available off the shelf but didn't like any of them.

"They had architectures based on minicomputers of the 1970s, which had shown how to scale a machine up from quite small, to significant computing power, as demonstrated by the successful Digital VAX-11/780 – the original one-MIPS computer, one million instructions per second. The 16-bit microprocessors were trying to emulate that growth path. But we saw two big drawbacks: real-time response to interrupts and memory access speed.

"When a processor is interrupted by an external request, how long does it take to respond? We used interrupts extensively on the BBC Micro whose 6502-processor had good response. Others with complex instruction sets were much slower to respond. Consider the NS32016: Its memory-to-memory divide instruction takes 60μs to complete 360 clock cycles at a clock speed of 6MHz. During that time, the processor cannot be interrupted. A double density floppy disk delivers a byte every 32μs, so you need an expensive direct memory access (DMA) controller for what is a simple task. This is the wrong answer in a cost-sensitive consumer market.

"After a lot of testing on different microprocessors, we found that performance was determined by how fast memory can be accessed. We found no performance advantage comparing one instruction set design to another. So, even though the NS32016 had a 'nice instruction set', and the 6502 had a rather simple 8-bit one, it made no difference to performance. It was these two factors: poor real-time response and poor use of memory that led us to look elsewhere for a solution.

"We were scratching our heads about this problem when some papers from Berkeley and Stanford Universities were dropped on our desks. Dave Ditzel and Dave Patterson were expounding the virtues of what they called the reduced instruction set computer (RISC). Their thesis was that the minicomputer-style architecture was the wrong answer. If the goal was to put the whole processor on a chip, the number of transistors needed to implement a complex instruction set didn't leave enough transistors to do more important things.

"Complex instruction set computers (CISC) have single instructions like: Enter a subroutine and save all registers on the stack, and perhaps do some other complicated tasks. The RISC processor was quite different. To perform a branch instruction to enter a subroutine, first save the return address, then set the Program Counter (also called the Instruction Address Register) to the address of the subroutine. That's all this instruction would do. If you need to save the contents of registers, to get space to use in the subroutine, that would be another instruction, or several instructions if you need to push several registers on the stack. So, instead of being a single high-level complex instruction, the subroutine entry would be broken down into a sequence of small simple instructions.

"It sounds like it's less efficient and is why complex instruction set micros were flavour of the day. The sales blurb called it 'reducing the semantic gap between the instruction set and the high-level language'. Meanwhile the RISC processors were going the other way. They were 'increasing the semantic gap'. But the argument was: If you keep it simple, you have more transistors available for things that are more important – things such as pipelining the execution code, which has the biggest impact on RISC performance.

"Without pipelining, the instruction is performed by fetching, decoding and executing the instruction, storing the result, then moving on to fetch the next instruction. With a pipeline, it fetches an instruction, then while you are decoding it, it fetches the next one. Then, while it is executing the first, it decodes the second and fetches a third. So effectively, all these resources, which are separate bits of hardware on the processor, are being used at the same time on different instructions. Instead of using them sequentially, they're used concurrently, which increases throughput by a factor of three or four times.

"In 1983, we decided to design our own microprocessor using RISC concepts. Guided by those published papers, Sophie started with the instruction set design. We were encouraged by Hermann, who saw this as an opportunity. He's not a computer architect, but he recognised a good idea when he saw one. He knew us well enough to trust our judgement. If we had pitched the idea to almost anyone else, I don't think we would have been taken seriously. Making a DIY microprocessor was a huge gamble, particularly with people who had never done it before. He thought it was an interesting way to go. He backed us, and it worked.

"Sophie designed the instruction set architecture and I looked after the microarchitecture. I think it was Andy Hopper who advised Acorn to invest in silicon design tools, and silicon designers. We identified what we thought was the best silicon design software (from VLSI Technology in California) and we recruited some experienced chip designers. At first, we didn't have much for them to do. They designed the second-generation Electron chip and one or two other things, but we had a capability that was under-utilised. When I finally completed the micro-architecture design, they were ready to do the silicon implementation. Eighteen months later, on 26th April 1985, we had the first working ARM chips back from VLSI.

"Unless you are someone like Intel or Texas Instruments, you don't make the chips yourself. We needed a fabrication plant (fab). We didn't have a billion dollars lying around to build a fab, nor did we want to get into the business of selling chips in the high street or to other hardware developers. Acorn was not a classic semiconductor company that designs a chip, fabricates it, then sells the device into the open market. It was building desktop computer products and was developing these chips to go into its own products, not as components for sale. There was never any question of Acorn building its own fab. The foundry business was emerging by then, so the business model was not new.

"The BBC Micro had two semi-custom chips – fabricated as ULAs by Ferranti in Oldham. We had problems with those, so we looked for an alternate source, which was our first contact with VLSI Technology. We followed a similar path for the Electron ULA, so we had done chip design; but the ARM was our first attempt at designing the processor itself.

"But then, the wheels came off the wagon, and Acorn was taken over by Olivetti. Acorn had over-extended itself in attempting to break into the US market, and in delivering the Electron late. It was the sort of product that would be a filler for the Christmas stocking. We couldn't make enough for Christmas 1983 because of problems with the ULA yield. By Christmas 1984, when we made huge numbers of them, demand from consumers had fallen sharply. The moment and the opportunity had passed. Both of those events cost the company a lot of money. I think they were equally damaging, so Acorn was effectively bust.

"Inside and outside the company, the RISC processor was a closely held secret. During negotiations with Olivetti, we were not allowed to talk to them about it. When they agreed to buy and rescue the company, we told them; but they had no idea what it meant, nor how significant it was. To be honest, none of us imagined it would become so significant.

"They allowed Acorn to do this bizarre thing, with their own processor in their own products. ARM-based machines, starting with Archimedes, first shipped in 1987. Meanwhile, Olivetti kept its distance. To them the hardware design, the processor and the operating system were all non-standard – that is to say: not IBM compatible. They made Intel-based PCs. From their perspective, that was all that people wanted. They didn't know what Acorn was doing; but fortunately, they let us get on with it.

"In the early 1980s, Acorn's business grew exponentially on the back of sales to UK schools, but then it went flat. Each year the government declared its budget for computers in schools, and that was Acorn's business. They didn't get all of that, but it set a cap on sales, which stopped growing. It was becoming more difficult to maintain processor development, which was getting more expensive to keep competitive.

"By August 1990, I felt my involvement had become less critical, so I left to take up the ICL Chair of Computing at the University of Manchester. Very soon after that, Apple approached Acorn about using the ARM in their Newton products. ARM was already in Apple products through a Radius graphics accelerator card. VLSI Technology had licensed the processor design from Acorn and had sold it to third parties. There was about to be more contact between ARM and Apple. The Newton was in trouble as it used the AT&T Hobbit processor – an interesting chip (optimised to process C compiled programs) but sluggish in performance. Interestingly, Hermann started the Active Book Company with ARM, then switched to the EO Hobbit chip. He went the other way; but that's another story.

"John Sculley at Apple wanted to use the ARM chip in the Newton but had a problem being associated with a competitor's name. They approached Acorn which proposed setting up a joint venture. They were received with open arms. In my last two years at Acorn, I spent quite a lot of time trying to find ways to set up the processor activity as a separate business, so lots of the groundwork had been done. We just couldn't find a model that made business sense. But when Apple came, they were open to the idea of a joint venture.

"By November that year Advanced RISC Machines Limited was set up in an old barn in Swaffham Bulbeck, and the processor was in independent hands. Its first CEO, Robin Saxby, introduced the business model that made it all work. He found the trick that we had failed to find. The problem was that standard licensing is a royalty business, but royalties come very slowly, and downstream. Robin introduced the 'join the club' model, where you pay your licence fee up front, quite a big licence fee. Royalties are terrible for cash flow, until sales have really built up. The upfront licence fee is good for cash flow. It's a big chunk of money, very early in the engagement. Robin said, 'Over my dead body will we make those chips. We're only going to design them.' I don't think ARM Limited ever seriously thought about manufacturing."

When things don't go to plan

"It's an occupational hazard for designers. We all like to think that we can catch every bug before we ship our first widget, but bugs and errors are inevitable. My biggest error in the BBC Micro was *disk error 14*. Going back to the Acorn System 3, I built the first floppy disk interface using the Intel 8271 floppy disk controller chip at home in 1979 or 1980. I studied the datasheet, set up the parameters, and got it to work. After some testing, I took it to Acorn, 'Look, I've got a floppy disk interface here.' And before you could say 'flux reversal', it was an Acorn product.

"Fast forward to 1981 when John Cox developed the BBC Micro's disk filing system. With hundreds, possibly thousands sold in Systems 3, 4 and 5, John assumed that my design was bug free. Five years later, with hundreds of thousands in the field, *disk error 14* was reported a lot more than we'd have liked. After some investigation it was found to be of my making; because when I set up the parameters, I got one of them wrong. No, I don't feel guilty – I just knocked up a prototype in my back room.

"Most of us agree that Acorn achieved what it did as a company, and we achieved what we did as individuals because we were incredibly lucky most of the time. Sometimes we took too much for granted. Even with talented and committed people, quality management systems are needed in every part of the business for the times when luck runs out. For Acorn, that lesson was learned too late.

"As for me: I've retired from full-time engagement at the University, but I shall continue to contribute where I can for the foreseeable future."

00001100

Sophie Wilson

Sophie Wilson is said to be one of the most influential women in computing in the world today. She was appointed Commander of the Order of the British Empire (CBE) for services to computing and elected a Fellow of the Royal Academy of Engineering and a Fellow of the Royal Society. She is an honorary fellow of Selwyn College, Cambridge, and has received the Royal Society Mullard Award with Steve Furber for work on ARM and was honoured as a Distinguished Fellow of the British Computer Society. In 2022, she shared the Charles Stark Draper Prize for Engineering with Steve Furber, John Hennessy and David Patterson for their invention, development and implementation of RISC chips. This is Sophie's story…

"I was born in Leeds and raised near Harrogate, though I never managed to cultivate a Yorkshire accent. I'm one of three kids. Our parents were both teachers. My father taught English, and my mother, Physics. Their only career advice was: 'Don't become a teacher.' We took mathematics at university. Physics and English didn't get a look in.

"We often helped them make things like boats and furniture. If my mother needed instruments for school, we built them too. The whole family would often sit around the table in the evening building things for mum like Heathkit multimeters. So, we learnt how to make things work.

"There were long summer holidays when we camped on Honister Pass in the Lake District. We loved to read so we went to the library each week and drew out four books each to share. We understood basic electrical principles but couldn't get any electronics books to guide us.

"In grammar school I was in the middle range. When I took my O Level exams, I was just average. For my A Levels, I chose Further Maths and Physics which put me into a class with only six other pupils. I did well enough that I applied for the University of Cambridge. After four months of intensive study, I completed the entrance exams, was interviewed and accepted. No one could have been more surprised than I was.

"For the rest of my gap year, I went to work for ICI Fibre Research in a small department that supported machines on the production line. Those evenings sitting around the table building lab instruments paid off. They asked me to build a 'wrap detector' to alert them if the fibre being fed to the machine broke. The problem didn't seem difficult, so I spent the morning building something that worked. I got a stern talking to by the union steward who said, 'Look, the time to build something like that is a few weeks. You were not supposed to build it in half a day.' He had a word with the boss. To avoid an industrial dispute, they moved me.

"They wanted other instruments built; one was a machine that counted droplets to maintain some subtle chemistry in the fibre-making process. The best way to control the concentration was to put the chemicals in drop-by-drop. They wanted an instrument using an infrared beam to count them. The detector counted each falling droplet as it interrupted the beam. I knew enough about electronics by this time to design and build something that would work.

"This was 1976. ICI used logic dating back to the 1950s. I had read a 1972 RCA CMOS data book from cover to cover. I showed the book to my supervisor. 'This works off 15 volts, has high noise immunity, and draws very low power, so it won't get hot.' They let me build the drop counter with CMOS circuitry. I later used a lot of Verowire and wire-wrap at Acorn, applying skills I gained on that project. After I got it working, I built another wrap detector; this time using CMOS and opto-detectors. That was about it for me. I left for university in September. They decided to miniaturise more of their instruments using CMOS. It was the first time I'd been able to influence the future – and I liked the feeling.

"At the end of first year, I joined the Cambridge University Computer Society and later the Cambridge University Processor Group (CUPG).

"I came home for the break and got another holiday job. One of the people who had worked at ICI wanted me to build an electronic cow feeder. I used a 6502 microprocessor to dispense the feed, manually controlled by the farmer, or triggered by tags that the cows wore. Being a farm environment, it had to be waterproof, and the buttons had to be reliable. We use the term *human machine interface* now, but if it existed back then, I hadn't heard of it. I wrote the control program for it and got it working. I first learnt programming at ICI. Here was I, a year later, writing machine code for a cow feeder and designing a computer!

"I went back to university and failed second year maths, which meant I had to see the Senior Tutor and try to work out what I could do next. By then I'd joined the CUPG where I told people about the 6502 processor, low power CMOS design, and generally made a nuisance of myself. People had their preferences, but I may have been more opinionated than most. For better or worse, people got to hear about me – in particular, Hermann Hauser.

"Hermann doesn't remember appointments, so he wanted an electronic diary – what we now call a personal digital assistant, as you would find on your smart phone. Well ahead of his time, he wanted one in 1978. He was a physics student, and a very bright one at that. He knew it had to be low power and here was I, talking about low power electronics. He contacted me and asked, 'I want to have an electronic pocketbook. Can you tell me how to go about it?'

"Without giving much thought to the demand on my time, I said, 'Oh, I can build one of those – no trouble. What you want it to do, and I'll design it.' So, we talked, and I went off to think about it. Two or three weeks later, I showed him my design. 'Look, this is what I have for you. This is how the keyboard works. This is how the display works,' and so on.

"He looked at my other drawings, then returned to the pocketbook and asked, 'Will this work?' Without thinking about how much time it would take, 'Of course it will,' I replied confidently. 'It is mostly just the main part of my own computer.' He asked me to build him one.

"Hermann and his friend Christopher Curry had set up Cambridge Processor Unit (CPU) – an Austrian's idea of a joke. They wanted to build an electronics consultancy business and had a job with ACE Coin Equipment that made fruit machines. Steve Furber and Chris Turner built a system using two SC/MP microprocessors. ACE reported that people were using piezo electric lighters to trick machines to pay out. They needed a way to protect the circuitry – to fix this problem.

"They got me to build some circuitry to prevent it. Christopher had a small Sinclair radio receiver that I could use to detect abnormal electromagnetic activity. When my detector was triggered by an electric lighter, it shut down the microcontroller, by asserting the 6502's reset input. It worked well in our trials. During acceptance testing, they plugged an arc welder into the socket next to it and started striking electric arcs. The machine was perfectly happy to shut itself down and say 'No, no, you're trying to fool me.' So, it passed the test.

"With all my extracurricular activities, I failed second year Maths and had to beg the Senior Tutor to allow me to take another subject. I was offered a one-year course in computer science. Reluctantly, I took that subject, though I didn't think of computing as something I wanted to do.

"For the rest of that summer, I worked late most nights for Hermann. I built my computer, which I nicknamed the *HAWK*. I don't recall why. He decided to commercialise it. That became the Acorn Microcomputer and the System 1. I had to change the design a little to suit the parts they had available, but it still used the 6502 processor.

"Chris Turner – CBT to everyone who knew him – ensured we could build and support the computers we designed. Chris takes and keeps notes like no one I've ever known. Totally practical, he has an eye for detail, and a remarkable memory. When I showed him my ideas for the *HAWK*, he came back with a neatly drafted design. It had just enough of the *HAWK* in it to avoid hurting my feelings.

"The prototype was made with Verowire, using the 160 mm Eurocard format, a seven-segment display and the INS8154 RAM I/O chip that Christopher obtained from Science of Cambridge stock. I wrote 512 bytes of machine code for the monitor. That doesn't sound like much, but it was written by hand with no debugging tools. I blew the PROMs and plugged them in. After I found a couple of minor bugs, it worked. That's how Acorn became a manufacturer.

"It would be untrue to say I was studying for my degree. I spent most of my time building things for Hermann. Any spare time I had was spent completing my studies and assignments. Before my final exams my parents came to talk to Hermann. Together, they decided my future. I didn't have much to do with what went on. I was still quite unworldly. Hermann took them out for cream tea in Grantchester on the river. They talked money. I was to be paid £1200 a year, and if I built a computer out of Acorn parts, I could keep it. (My parents weren't great negotiators.) I guess it was a bit unusual, but that's what happened.

"It was the summer of 1979 when I started full time at Market Hill. Our priority was to develop add-ons for the System 1. It was a huge task to bootstrap the business. We had nothing, so I wrote code by hand. I'd write out the mnemonics, hand-assemble them, then enter the hexadecimal code into a PROM programmer. I wrote a 6502 assembler as soon as I had a working machine. It was about four kilobytes of code, which also had to be written by hand. It was a really, slow process, but it was the last thing I had to write by hand. Happy with the progress I'd made, I went home for the Christmas break and wrote my first BASIC interpreter.

"Until BASIC became available, the first System 1s were bought for the same reasons that people bought the Sinclair MK14 – as a hobbyist machine for writing machine code. I designed everything except for the cassette interface, which Steve did. I wrote the software, wrote the user manual, designed the circuits. Chris Turner wrote the technical manual. When orders came in, I helped pack the components, shipped them, and answered telephone calls from people who couldn't get them to work.

"There were two lines of business activity going on at the time: consulting (like building fruit machines for ACE), and retail sales of the System 1. Everyone pitched in: Hermann, Chris Turner, and when they were around, Christopher and Steve would help too. It was some time before we had specific roles that would allow us to concentrate on just one part of the business; but even when we did, it was never beneath any of us to help when needed.

"The monitor firmware allowed the System 1 to be programmed in hexadecimal, so users were using it much as I did before I wrote the assembler and BASIC interpreter. They'd save their work on cassette, rather than having to enter the program each time they wanted to run it.

"Because of the consultancy business, we designed card frames to take extra hardware. The System 1 would plug into a backplane, then we developed all sorts of other modules to connect external devices. There was an eight-kilobyte memory card. I think we built three different types of video display, an in-circuit emulator, a programmable input/output card, a disk interface, and an HPIB/GPIB bus controller – everything you'd expect from a company making control systems. Whatever the assignment, we would assemble a development system out of those modules. Once we had a working prototype, we'd often build a dedicated board to combine functions the customer wanted onto a single board.

"It seems that 1979 was the year of the Eurocard System for the motley crew at Market Hill. Steve and I did a lot of design work, with input from Chris Turner, and later Laurence Hardwick, Paul Bond, Kim Spence-Jones, Joe Dunn, Carl Dellar, and others. We would see a need, and if we didn't have the people hanging around, Hermann would talk to his network of contacts at the university and would find them. He had a flare for finding the right person for whatever job we needed done.

"Acorn sold modules to hobbyists who wanted to build the same sorts of machines. A System 1 plus a card frame made a System 2. Add a floppy disk controller and disk drive made a System 3. A System 4 had two racks of card frames. The System 5 was an improvement on the System 4, with the 6502-processor running twice as fast. There were lots of parts to keep track of, then Colin Priestley joined us to help Chris.

"Finally, Christopher left Science of Cambridge and joined Acorn full time. He wanted to build a computer for the home hobbyist – something that looked less 'industrial' than Systems in Eurocard racks and it had to be more affordable. He and Nicholas Toop built a small computer out of parts from our store. It used a lot of the System hardware and software which saved time and simplified stock control.

"Christopher and Nick disappeared into the Fens for a while and when they reappeared, they had re-packaged the System 2 into a small plastic case. Whatever you might think of the Atom as a computer, it was marvellous as a piece of industrial design. Allen Boothroyd was brilliant to work with – always creative, practical, enthusiastic – he continued to show those qualities over the years.

"The Atom had technical faults like a video chip that could only support the American NTSC television standard. When plugged into a UK TV set, it might or might not work. Also, the motherboard was built with its chips fitted upside-down. So, if the keyboard was pressed too hard, the chips would fall out of their sockets and lie in the bottom of the case. There were problems with it, but it sold in large numbers. What it had in its favour were features inherited from the Eurocard Systems. A lot of their flexibility was included by default; and when Paul Bond, Carl Dellar and Joe Dunn added Econet, Cinderella was ready to go to the ball!

"Hobbyists were buying Atoms, lots of them. It seems hard to believe but 12,000 were sold. That was enough to whet Christopher's and Hermann's appetites for something better suited to the home computer market.

"After writing the first version of Acorn BASIC it was used everywhere. There was a version with floating point added for the Atom. Acorn's accountant even engaged Graham Tebby to write a full accounting system using Acorn BASIC. With each evolution, I refined my ideas about what should be in it.

"By December 1980, with Systems and the Atom shipping in large numbers, we started thinking about what would come next. By early January, we had a lot of ideas, but nothing firmed up. Where to take the company was always a topic of discussion but around that time, we seemed to have even more spirited debates. Okay… Heated might be a better description. Andy Hopper wanted us to build workstations, using a 68000 processor, or similar. Christopher and Hermann wanted to build something more like an Atom."

The Tube

"I proposed building a machine that had two processors, but could be cut in half. People could just buy the Input/Output section to use as a home computer, which we called the Proton. Communication between the two processors would not be the same as a data or address bus, which are internal connections in computers. This did a similar job but in different ways. So, what would we call it? According to legend, Steve was traveling on the London Underground when it came to him. He was on 'The Tube' which was clearly 'not a bus'. So that's how the second processor connection got its name.

"Acorn had become a successful manufacturer – the British equivalent of Apple. The Proton, like the Apple II and a few other microcomputers of that time, would run a 6502, though much faster, had higher resolution graphics, and more inbuilt connections to the outside world. We employed a lot of Eurocard System capabilities in the design, but the wire wrap prototype built in the first week of February 1981 was only a proof of concept.

"After the BBC came to call and it was announced a week later that Acorn would build the BBC Microcomputer, the rush was on to complete the design and get it into production. That year was a blur for me. BBC BASIC occupied all my time and attention. But in the end, I feel we came up with the best compromise between what I believed was important and what Richard Russell and the BBC wanted. Considering the complexity of the task, it's had surprisingly few revisions over the years and proved its worth time and again.

"During early development, people within Acorn and the BBC had different ideas about what we should build: a small cheap machine or a big expensive workstation. Developing the video and serial ULAs had forced us to learn how to build custom integrated circuits. We would need another ULA to do the heavy lifting for the Tube interface between the Beeb and the second processor.

"I designed the Tube interface architecture and wrote the software protocols that ran it. Steve developed the circuit design and produced the artwork for the Tube ULA. It was a bi-directional direct memory access (DMA) controller with first-in-first-out (FIFO) buffers to stack up data when things got busy. The operating system in the second processor could send commands and data across the Tube, and the I/O processor would execute them.

"The I/O processor was a very capable self-contained computer. That's what the BBC franchised as the BBC Model B. Acorn and a few third-party developers would eventually implement second processors for most popular microprocessors; among them were the 6502, Z80, NS32016, 80186, 80286, 6809, 68000, 68008, 65C102, and ARM.

"With the exception of the 6502 and ARM second processors, Steve and I were unimpressed with the performance of these chips. For a start, they didn't make good use of memory. Secondly, they weren't as fast as their advertising hype claimed. Finally, they weren't easy to use. We were used to programming the 6502 in machine code and we rather hoped we could get the same or better results writing in a high-level language. For that, we'd need one of these so-called 'more powerful' microprocessors.

"We wanted to build machines that could run 3D graphic games without having to write assembly language programs, but processors were too slow. We needed more processing power to compete with IBM PCs and compatibles, which were beginning to flood the market. Considering the wider bus, more powerful instruction set and faster clock speeds, our expectations seemed reasonable; but what we got was far from what we expected. We felt we needed a faster and more efficient processor.

"The only second processor that sold well was our 4MHz 6502. When linked to a 2MHz 6502 I/O processor (a Model B), everything ran seamlessly, and programs loaded and executed twice as fast; and when handling a lot of input/output operations it appeared to run even faster due to the I/O processor taking over some of the work.

"Some people wrote games for the dual processor, like David Braben's and Ian Bell's game, *Elite*. The second-processor version of this three-dimensional game showed off the processing power. Second processors with 16-bit CPUs (like the 8MHz 68000 and 80286) were disappointing – no better than the 4MHz 6502. Hermann had to rethink his ideas about selling a multitasking Acorn business machine running the NS32016.

"Processors with double the clock speed and buses twice as wide were often slower due to inefficient use of system memory. On these other micros, memory access takes four cycles of the master clock. An 8MHz 68000 would only access its two-byte-wide bus twice per microsecond; so, their bandwidth was only four megabytes per second, just like the 6502. We could have designed a 16- or 32-bit-wide high-performance memory system with a bandwidth of 16–32 megabytes per second if we could find a microprocessor that behaved like the 6502. We became convinced, and we still think it's true for cache-less microprocessors, that memory bandwidth is an indicator of system performance. In other words: We could build a faster memory system, but we couldn't find a microprocessor with matching speed.

"We visited the Western Design Center (WDC) in Phoenix, Arizona. They were working on a 16-bit successor to the 6502 – the 65SC816. Andrew McKernan (head of Acorn Advanced Research), Steve and I visited William D. Mensch, who had co-created the 6502 with Chuck Peddle. We had already been to Israel to talk to National Semiconductor about their NS32016 development. They had large teams working in spacious facilities within a science park. But at WDC, we saw a small team working in bungalows, packed in tight and using ordinary drawing boards. There were some college kids designing processors by sticking strips of Rubylith tape on film. They were even less sophisticated than we were! Although the 65SC816 was an advance on the 6502, we decided that it wasn't the answer.

"We realized that if they could build a processor, so could we. That's when we started thinking about designing our own. Our project slogan became 'MIPS for the masses' – millions of instructions per second for everyone. Few manufacturers of microprocessors shared our thinking. One day, Andy Hopper dropped press releases and papers on my desk that described *Berkeley RISC* and *Stanford MIPS*. They introduced me to RISC technology. A month later the IBM801 papers were published. I studied them all, until I could recite every idea they contained. Steve, Hermann and I talked about building our own RISC processor, and what we would want it to achieve. I argued, 'We can do this. We can get the MIPS we need.' Hermann was an integral part of these discussions, bouncing ideas and suggesting how we should go about it.

"I'd designed an instruction set at university. I built on that experience. Steve researched processor design – things like how a pipeline works. We decided that the 6502's semi-pipelined structure was more efficient than what Motorola did in the 6800 and 6809. We liked to see things overlap well, so we decided we would pipeline the memory BUS too. It was later removed because people found it too confusing.

"Hermann, Steve and I worked separately in the mornings, then we'd go down to *The Sportsman* or *The Rose & Crown* for lunch, talking about what we were doing; then we'd come back and revise what we'd discussed. We became increasingly convinced that we could make this thing. Hermann set up an official project in Acorn. We started with Steve, Hermann and me. Gradually, we recruited four VLSI design engineers, initially led by Robert Heaton, and later by Jamie Urquhart."

Our Game Plan

"An Acorn RISC Machine (ARM) chip would need to be 'quite powerful' but not 'super powerful'. That may seem strange to say now, but Steve and I knew what we meant. In time, we got others to understand it too. We'd endowed it with the ability to perform cheap but powerful computing. Making it very low power was a pleasant surprise. In hindsight, it should have been obvious. We only had 25,000 transistors in the ARM1. We were worried about power dissipation. We needed to be extremely careful – something that could be mass produced and fitted to cheap machines without heat sinks or special cooling. So, there were already aspects of power conservation in the design, but we achieved far better than we imagined; and as the world has become increasingly more mobile, that aspect of ARM has mattered.

"We came to understand that while others were adding large amounts of processing, ARM offered a different approach that jumped over Intel and other manufacturers. In striving for what we thought would be ideal for our market, we designed a power-efficient embedded processor. Not our intention; but in the end, that was a side effect of making them cheap and simple to use. People may call it just dumb luck, but it was OUR dumb luck! It would be more charitable to call it serendipity."

No dead end

"We did have some doubts. We kept looking for the intractable problem that meant we wouldn't be able to do it, but we found no dead ends. It was just a matter of doing the work. We gradually converted Acorn's Advanced Research and Development department into a dedicated group of people working on the project. After we realized the project would succeed, we decided to build the input/output (IOC), video (VIDC), and memory (MEMC) controllers as well as the microprocessor (ARM). So, we built four chips. Because we designed it first, the ARM processor arrived about six months ahead of the rest.

"Mike Muller and Tudor Brown (who later joined ARM Ltd) designed the IOC and VIDC chips. Alistair Thomas (who sadly, has since died) built the memory controller with Steve. Steve and I built the ARM processor. There wasn't any major insight or breakthrough. We just read the papers, absorbed what was required, and went through the design process. There weren't any great insights or *Eureka* moments. But we did it!

"All the chips we built worked first time. All but one went into volume production as Revision A. There was nothing magical about it. We supported each other: 'We can do this. We just need to get on and do it.' We expected to find roadblocks. We expected to discover why National Semiconductor employed so many people in Israel, and why they found it so difficult; but we never did. We just thought it was going to be much harder than it turned out to be.

"We put a lot of effort into verification and modelling. Steve wrote the first behavioural model of the ARM in BBC BASIC. We ran test programs on that. I led a group who wrote test programs to check the behaviour of each instruction. We also wrote instruction simulators so we could write more software. Being written in BBC BASIC, the behavioural model was slow, only a few cycles per second, but it worked.

"It was different with pure instruction simulators, where we only wanted to check the instruction stream. Timing and hardware performance was not of concern. Here, we could run at hundreds of thousands of ARM instructions per second on a 6502 second processor. We could also write a large amount of software, port BBC BASIC to the ARM and anything else that was needed – like a second processor operating system.

"With this work, our confidence grew. We were getting results better than anything else we'd seen. Although we were interpreting ARM machine code, the result was often better than compiled code on the same platform. That was really encouraging.

"We sent the ARM design off to VLSI Technology to be fabricated. It came back on the 26th of April 1985. We plugged it into the ARM second processor board that was ready to test the chip. The Tube operating system booted up. It ran BBC BASIC. We said, 'Print Pi!' and cracked open bottles of champagne, because it worked.

"I don't mean to seem smug, chip design wasn't a great struggle as others found it to be. For example, I believe the NS32016 was into Rev J at that time, and it still wasn't right. Our team had fewer than 10 people. Perhaps keeping the group small was part of our success.

"We did a lot of things in parallel. Hugo Tyson, Jes Wills, Jon Thackray, David Seal (another great person who is missed) and I were writing verification and validation software and interpreters. A team led by Lee Smith and Harry Meekings wrote compilers for it. There were others, but making the processor itself needed a small number of people.

"We sold the ARM1 as a second processor development system. There were lots of people eager to get their hands on all that computing power – particularly people in universities. Researchers quickly produced a lot of software for it. We had Prolog and LISP. We hadn't managed to get Acornsoft to develop a C compiler for the BBC Micro, but in short order, we had C and BCPL compilers from university people for the ARM.

"When all the chips were qualified, we produced a complete Acorn machine. It didn't have a great operating system at that stage. Though *Arthur* was just a scaled-up version of the BBC Micro operating system, it allowed us to sell machines, while we developed RISC OS – which proved to be quite effective. Even on its first release, we had some world firsts: In RISC OS we had anti-aliased typefaces designed from hinted glyphs, all computed at one time, and that was five or six years before the Apple Macintosh or anything else would do that. We had much more computer power than they did - and the motivation to get it done."

Becoming an Embedded Processor

"By about 1988, we had plans for a new processor that Acorn couldn't afford and looked for ways to fund development. It took three attempts to put together a viable business model. We began to get feedback that people wanted the processor free of any Acorn connection. Apple was one of the companies that wanted to use the ARM chip, but they also wanted to keep their distance from Acorn, as we controlled the whole thing, and were seen as their competitor; so they supported the notion that ARM could be taken out of Acorn to become its own entity with its own resources, jointly controlled by Acorn, Apple and VLSI Technology. So, the founders of that ARM startup would control the destiny of the processor, and each would also be a customer.

"In November 1990, Widelogic Limited was incorporated as a joint venture. Acorn provided 12 employees (Jamie Urquhart, Mike Muller, Tudor Brown, Lee Smith, John Biggs, Harry Oldham, Dave Howard, Pete Harrod, Harry Meekings, Al Thomas, Andy Merritt, and David Seal) and the intellectual property, VLSI provided tools and Apple injected US$3 million. A month later, the name changed to Advanced RISC Machines Limited. It had some tough years before it became large enough to have the resources to give customers everything they wanted. Along the way, it also had to invent an effective business model.

"Robin Saxby came on board as the first CEO. Without that, it wouldn't have survived. But with that, people could build their own chips with much lower risk. Robin created the ARM ecosystem that exists today out of that business model – by licensing access to the architecture and the designs, then effectively selling a licence to use a design. The next step beyond fabless chip companies is design-less: You only give away the design. You don't get involved in making the chips at all.

"When Steve left Acorn to become a professor at the University of Manchester it was very difficult. Professionally speaking, Steve and I had been joined at the hip since we met at university. I had to make some adjustments. At work, I did a lot of things apart from designing ARM processors. I wrote software, designed end-user computers for Acorn, and the processors that went into them. I stayed with Acorn, because I felt that joining a small chip company would limit my freedom to take on projects. Acorn contracted my services as a consultant to ARM. They needed my help, particularly in the very early stages, until they could get it running themselves.

"At Acorn, I spent the 90s doing a bunch of things. I wrote a multimedia software system from scratch – Acorn Replay, Acorn's answer to Apple

QuickTime. We designed a video phone, and we kept designing new computers. Eventually the Internet came along, and people were attracted to the idea of having computers running over the Internet.

"Oracle wanted a very low-cost client computer. Windows machines cost ten times what Oracle wanted to spend. It could be done using Acorn's kit of parts, so we built what they needed. We developed an entire video-on-demand cable TV system for the local Cambridge cable company. There were all sorts of projects. We built some fascinating computers, and some equally fascinating systems on chips (SOC).

"We worked with the semiconductor division of Digital Equipment Corporation (known as DEC, then later just Digital) to build the StrongARM processor, and then its successor for multimedia, the SA1500 (a StrongARM with a media processor) and the SA1501 (a single-chip I/O system with memory and video controllers). Those chips and the memory array made a complete software-programmable set-top box. It was a powerful system that could decode two MPEG-1 streams – one picture-in-picture, all decoded in software. When Intel bought DEC, along with Digital Semiconductor, they didn't know what to do with that project, so they killed it. C'est la Vie!"

The End of Acorn

"When ARM Ltd was floated on the London Stock Exchange in 1998, there was finally a way to establish the real value of Acorn shares. As ARM became more successful, its share price went through the roof. Acorn's shareholders could benefit from ARM's success. By June 1999, its ARM shares were valued at more than £127 million. Since the future value of the business lay with ARM, not Acorn, Morgan Stanley proposed a way for shareholders to exchange their Acorn shares for ARM shares. For me, as an employee of Acorn, the writing was on the wall.

"If Acorn ceased to exist as a manufacturer I would need a new job. I'd been developing another microprocessor, code named ALARM (A Long ARM). I was spending my time thinking about 64-bit buses, 64-bit instructions, and so on. People inside Acorn thought, 'Hmm, what could we do with that? Could we start a company with ALARM as its seed – a processor that was substantially more capable than ARM.' They hawked this idea around Silicon Valley venture capitalists, but the answer was always 'No'. They wouldn't fund another company like ARM with the same idea. They didn't see much return in it for them. After several rejections, they found a venture capitalist (VC) who said, 'You need a business model with a much shorter exit time for the VC, because if you're licensing IP as

Arm does, the time to get a payback is long. You've got to take some of those steps out, so VCs see a return in their timeline. You've got to be the people who make the SOC that exploits your intellectual property. You should become a fabless semiconductor company. I know you've had these ideas about media processors and digital TV, but I don't think that will fly. You ought to get into communications. I think you ought to build something on this new-fangled ADSL standard.'

"Well, it was G.Lite at the time – before ADSL. So, we adjusted our business plan and got funding. ALARM was renamed FirePath, and we found a silicon implementation team from INMOS/ST in Bristol, and a modem team from Alcatel at Mechelen, Belgium. We put together a three-site startup called Element 14 that exploited the FirePath processor. We started off calling it G.Lite, but FirePath was far more capable than that. When the ADSL-1 standard came out, we said, 'Right, it can do that.' In fact, it could do six ADSL-1 modems per FirePath, and we put together multi-core chips. We were a multi-core chip company from day-one.

"Within about a year of leaving Acorn, Broadcom bought us after just one funding round. Broadcom is the volume leader by a long way in central office DSL modems with FirePath processors. Now I work for a chip company, designing chips – just what I said I wouldn't do.

"I count myself as being very lucky. I worked on two wildly successful microprocessor designs. At Broadcom I continue to work on similar things. Do I have another microprocessor in me? Well, FirePath is set up on a different basis to most processors. It's much more flexible and FirePath has been changing as a result. The program I wrote in 1983 as the first demonstration of what an ARM could do still runs on a Cortex A15 today. FirePath isn't built like that; it's morphing faster than anything I've ever seen and changing radically as a processor.

"I'm proud of every project I've been part of and grateful for the opportunities that came along. I've had the pleasure of seeing things that I design go out and be used in volume. Back in 2008 when I was first told they'd shipped 10 billion ARM cores, I was impressed. Now they've shipped over 300 billion and are shipping 30 billion a year. Who cares about that anymore? But I still have the satisfaction of seeing my ideas go into things that are made to work. I suppose that goes right back to my parents who taught me how to do just that.

"It's hard to digest. I mean, there's about 20 billion ARM7s in use. To have 20 billion copies of a little bit of my brain running around the world is scary at times. I sometimes wonder how much more I could have achieved if I'd paid more attention as an undergraduate."

00001101

Andrew Gordon

Andrew Gordon joined Acorn as the first shipments of BBC Micros were leaving the Cleartone plant in Wales. He spent his gap year working on its operating and filing systems, followed this with three years of study at the University of Cambridge, 20 years at SJ Research, and occasionally returned to Acorn for special projects. This is Andrew's story…

"My first exposure to Acorn Computers began in the summer of 1981. I occasionally visited their Market Hill office while completing my A-Levels at Felsted School in Essex, where we developed Econet and Prestel interfaces. On one visit, I unwittingly purloined a carload of Atoms, which left me with some valuable life lessons.

"I went to borrow computers for a conference. Unaware that my teacher forgot to arrange the loan with Christopher Curry, I drove to French's Mill and said to someone in the store, 'I'm from Felsted School, here to collect some Atoms.' They helped me load my car with all the machines it could carry; then I drove away. No one knew who I was, yet I didn't sign any paperwork. It was a great lesson in the need for stock control, security, and the dangers of making assumptions. The computers were all returned in good order, so I hadn't spoiled my copybook.

"Fast-forward six months to a cold, frosty Monday in December 1981. I had finished my A-Levels and was taking a gap year working at Acorn, before studying computer science at Cambridge.

"That morning, the south of England received an unusually heavy snow fall, but since it was my first job, I wasn't going to let a little snow stop me. I was determined to be there on time.

"Driving up from East London, the snow was deeper that I'd ever seen. Even on the motorway, driving was slow and nerve-wracking. Luckily, there was almost no traffic. I'd never driven in that kind of weather, so I was in no state to play dodgem cars with lorries. Once I left the motorway, I didn't see another car until I got to Cambridge. Roads were blocked so it was a crawl for the last part of the journey. Being one of the few stupid enough to be on the road that morning, I pushed through by driving on the other side of the road when snow piled up on my side.

"Arriving at Acorn's new offices in Cherry Hinton, I found the car park almost empty. It looked like I had come on the wrong day, but a few lights burning at the back of the building gave me hope that I wasn't alone. There was no one on the front desk. I waited, finally plucking up enough courage to wander into the offices. The recently renovated building was a far cry from the cramped conditions at Market Hill.

"There seemed to be at least one other soul because in the distance, I could hear a BBC News bulletin. I moved nervously in that direction, concerned that someone might think I was an intruder. I didn't want to be in trouble on my first day. The office was empty where a small radio blared out a weather report about dangerous driving conditions in the Southeast. The bulletin advised me and the rest of its listening audience to stay home. 'A bit late for that now,' I thought. But two offices down, I saw a familiar face, studying a set of galleys for the *BBC Microcomputer System User Guide*. It was John Coll. We'd met once or twice before.

"John jumped to his feet when he saw me, and with what I came to know as his genuine friendliness and enthusiasm, he welcomed me warmly, commented on the miserable weather, asked me if I had any problems getting to Cambridge, and offered me a cup of tea.

"He didn't know I was coming but phoned around until he found someone who knew what I was supposed to be doing. The BBC Micro Teletext Adapter had become a higher priority than the Prestel editing terminal I was hired to write, so that's where I'd be starting. I moved into a new office at Fulbourn Road, and the rest of the development team were in the process of moving there too. Within a few days, we were all settled into the new digs.

"John arranged a ride for me to French's Mill in a company van. I found the Teletext prototype and recall seeing a couple of one-armed-bandits lurking in a corner; left over from the Cambridge Processor Unit consulting job for ACE Machines. I hitched a ride back to Cherry Hinton. With the move to the Waterworks, things were a mess. There was no TV aerial at Fulbourn Road to connect the Teletext receiver, so it took a couple of days to fetch and install an antenna from Market Hill (which was in the process of being taken over by Acornsoft). I was feeling quite pleased with myself as I'd overcome everything Mother Nature had thrown at me and was still able to become productive – or so I thought.

"Later that first day, I met up with Christopher and Hermann. If they knew anything about the carload of Atom computers, they didn't show it – and probably wouldn't care if they did. I was fresh out of school, starting my first job, and had nowhere to live, so Herman said I could stay at his house while I found something more permanent. He was about to go to New Zealand, so a couple of days later, he left me to look after his house. He didn't know me. I'd been recommended by someone he also didn't know. That's the level of trust Hermann is known to show others.

"Except for Chris Turner, Sophie Wilson and Joe Dunn, there were few permanent staff in the development team. Laurence Hardwick was permanent, but he was setting up customer service. The rest of us were contractors or moonlighting while at university. Paul Bond worked for TopExpress, a Cambridge consulting firm. The reliance on contractors changed in 1982 when new people arrived to replace them. Paul was replaced by Jon Thackray as project leader for the Machine Operating System (MOS), though there never seemed to be any doubt that Paul owned the MOS in the same way that Sophie owned BBC BASIC.

"When the first BBC Micros shipped, they could only store and recall programs and data on cassette recorders. Disk storage and Econet were urgently needed, so the highest priority was to get the Disk Filing System (DFS) and Network Filing System (NFS) working. John Cox used Acorn DOS from the Eurocard System 3 to write the DFS.

"Teletext was a lower priority at the time, so I changed focus again and became the link within the Systems teams, sorting out details, and documenting decisions made about the interfaces between the MOS and filing systems. I wrote specifications, then developed the first sideways ROM as a proof of concept.

"The Systems teams were led by Paul. He wrote the first generation of the Econet's low-level code at Market Hill in 1980. It handled transmission and reception of data packets – the link between the operating system and the physical network (the chips, wires and clock). The NFS acts on requests from the MOS. Joe Dunn wrote the Econet client software which ran on Atoms and Systems. Carl Dellar wrote the first version of the file server and printer server after completing his PhD. He moved to the USA in the summer of 1981 after doing a great job setting the foundations for what Econet servers would become. Brian Robertson re-wrote the Econet low-level code, Jes Wills ported the client software which became the NFS, and Joe worked on the file server and managed the Econet team. Others helped Paul finish the MOS.

"Felsted School had been trumpeting the use of local area networking in education. In December 1980, while still in school, I was in the audience at a Regents Park hotel in London, when Hermann launched the Econet. Acorn showed Atoms running network clients. For the launch, Carl had developed what would become the Level 1 File Server. It was unfinished but performed well enough to impress everyone there. Few had ever seen a local area network – certainly none on low-cost microcomputers. It was my first product launch, and though it lacked the glitz of an Apple event, it was memorable for its technical achievement.

"Felsted School suggested connecting non-Acorn machines like the Research Machines 380Z (Acorn's main competitor) to Econet. Christopher Curry liked the idea, so we did it with resources supplied by Acorn. They gave Felsted some Atoms and a System 3 File Server to experiment, to show the use of Econet in schools, and to connect the RML machine. That's how I made my start with Econet development. Paul was the person I talked to when I had queries; and believe me, I was talking to Paul a lot.

"Fast-forward to 1982: I was helping to develop Econet for the BBC Microcomputer. The way it handled operating system commands was more advanced than the approach taken with Systems and the Atom. For example, if a *cat command (which lists a directory's contents) is executed in a Language ROM (e.g. BBC BASIC or VIEW), it is passed to the MOS, which sends the request to the active file system to action. This could be the NFS, DFS or if there is no other file system, the Cassette File System (CFS) in the MOS ROM – just like a 'real' operating system."

Acorn's coming of age

"It was common for us to take an interest in what others were doing. We all had our focus, but we turned our hands to any job that needed doing. For example, Brian Robertson and I spent evenings debugging the Econet cabling in the main building. We weren't doing it for extra pay or glory; we did it just because it needed to be done.

"In its consulting business model, Acorn took on almost any project. Someone would suggest an idea and if Hermann or Christopher felt it had merit, we would do it – no analysis, little planning, no costing.

"There were lots of other things going on. If something seemed like a good idea, Acorn was all for it. Christopher came up with the idea of having a Prestel-style display and editing terminal. He argued that they were quite expensive, and we could make them cheaper. There were lots of small projects going on before the BBC Micro. The Proton started as just another idea that a few people thought was worth pursuing. Later, when the BBC showed some interest, it was clear that this was the big one; so other jobs were now just background noise.

"1981/1982 was when Acorn flipped completely. It became a real company. Until then, it was very much a seat-of-the-pants operation – a juggling act. They had been doing more than they could manage using people from the university, contractors, and even getting people to do things for free. Winning the BBC contract was serious, and things changed quickly. They recruited a lot more staff. Rather than Christopher and Hermann running everything, they brought in four professional managers, starting in 1982. They were Jim Merriman (Manufacturing), John Horton (Technical), Peter O'Keeffe (Sales) and Peter Wynn (Finance)."

Lesson with Teletext

"While the contract was potentially very lucrative, the BBC was saying, 'We have a contract with deadlines you must meet.' That needed a new approach. The BBC was the opposite of us: hierarchical and formal. I remember going to a technical meeting in London that left a lasting impression. Some BBC people were talking to us about Teletext. There was Hermann. Mel Pullen designed the hardware and had most of the technical understanding. I had just started working for Acorn. Kim Spence Jones was there to lend moral support. I noticed that the BBC people paid most attention to Kim because he was wearing a suit. Hermann came next in their pecking order. He wasn't wearing a tie, but he was smartly dressed and had a PhD. I was wearing a jacket and tie, so I was shown some respect; and then there was Mel – our expert was mostly ignored because they saw him as being scruffily dressed. I was young and impressionable, but I remember it clearly because it was the first time I'd experienced that kind of prejudice.

"The Teletext Adapter was the first peripheral to reveal problems with the BBC Micro's 1 MHz Expansion Bus. To make matters worse, Teletext didn't work as described by the BBC. It was my job to resolve those sorts of problems. I had done some work on the software, but there were problems with the hardware that stopped me from getting on with it. With help from Kim, I debugged the analogue circuit. When the BBC's technical experts set it up, it worked fine, but if one of our engineers made the adjustments, it didn't work at all. Clearly, it was not usable as a production design. This was not an isolated example. Even with my limited experience, I could see that the BBC didn't understand their part in ensuring successful outcomes."

Keeping the ideas flowing

"Working with the operating system team before starting university was a brilliant way to learn my craft. At that point the MOS was still changing to work with the filing systems. I contributed to the project by compiling the revisions to the interface specifications. I can't claim credit for its development, but I documented it and contributed to the Teletext Adapter. John Cox developed the DFS and Joe wrote the NFS client software. With Paul's guidance, we pooled our requirements for what was needed and worked out how the MOS would manage it.

"Paul was supreme commander. In the case of the sideways ROM, I was tasked with finding out what each designer wanted and documenting all their requirements; then Paul reviewed them and often said something like, 'Don't do it that way, do it this way.' I think we all felt like we were doing an apprenticeship. He taught me more in three months than I learnt at Cambridge in three years. Paul expected to see beauty and order in software systems. While working on a proposal, Paul asked why I approached a task the way I did. I said, 'Yeah, I know it's wrong, but the other guys want it that way.' I lacked the intellectual courage to enforce good design principles. He saw that I was prepared to do what was technically 'OK' – settling for something that 'looked wrong'.

"Working there taught me a lot about doing things correctly. They called it 'good computer science'. You need to have all the theoretical principles in your toolkit, otherwise you build systems that don't work, aren't efficient, or are unreliable. I've since done some supervising at Cambridge; and it's good to see they now teach some of those ideas. They have an element in the software engineering course where students learn by working on big projects. I was lucky that my first job was a big project where things were done right."

Customer facing

"I think one risk to many development teams is becoming too isolated. When Acorn was small, it was in regular contact with customers, so it got to know what mattered firsthand. After winning the BBC contract, there was a Customer Support department downstairs and Development lost contact with the world – with one or two exceptions. One of these was Barson Computers in Australia, where the DFS and NFS were vital to their business. They were a year or two ahead of the UK, so they pushed us to perform. It was one of the few ways we got to hear what end users were experiencing. It was even better when Brian Cockburn joined us in Cambridge. 'Bruce' was the first member of the development team with field experience; and improved even more when Laurence joined us.

"The direction for new tasks usually came from Paul, with occasional input from Sophie. She was in the next office, but I don't recall much flowing between them. In the case of the DFS, it was clear what we needed to do because disk storage on Eurocard Systems was well established.

Paul's direction to John was: Take the code; don't change it; just make it work with the BBC Micro.' John worked with Paul to learn exactly what he wanted. Not touching the code needed less direction, but it proved to be a mistake when *Disk error 14* appeared. Steve Furber had written the code at home. It 'seemed' to work on System 3s, so it was never checked until about 100,000 BBC Micros with DFS had shipped with the bug. Another lesson: Don't assume. Check!

"The foundations of the MOS had been completed with Graham Tebby doing the graphics, Peter Miller writing the sound code, and Laurence looking after the cassette software. That was largely done and dusted before MOS 0.1 shipped, but there were other things still being worked on. Occasionally, other people wandered in to complete a task or adjust existing code in the MOS.

"When I left the project in August, I'd made some progress on the Teletext Adapter. I handed it over to Graham and barely looked at it again. My main contributions came from what I put into the MOS and my help with debugging. By then, Jon Thackray had become project leader, tying it all together, and taking requests from the minor projects."

SJ Research

"At university, I worked for Acorn on small jobs and at SJ Research, helping Kim build an Econet client for the 380Z, and later a file server. Research Machines dominated educational computing before the BBC Micro. Acorn saw the value in having their competitors' machines on Econet, so we developed an interface. Our customers were keen to install the software but said, 'We want to buy your product, but we can't get a clock box from Acorn.' Kim decided to make clock boxes and terminators – better than Acorn had done. He hadn't anticipated how profitable it would be to sell connectors, cables, clocks and terminators. Small things that weren't a priority for Acorn was quite profitable for SJ Research.

"It started with writing client code for RML machines, but some schools wanted to use their existing RMLs as file servers, so I wrote versions of the file and printer server. SJ sold RML machines, but then we made our own file servers with floppy and hard disk options. These developed a reputation as being the best Econet servers in the country, establishing SJ Research in education throughout their product life."

Electron panic attack

"By April 1983, Paul Bond had left Acorn. I was in my first year at Cambridge – then came the Electron panic attack. Steve Furber designed the Electron in the summer of 1982. Brian Jones wire wrapped different combinations of features to come up with an Electron that would be functional and affordable.

"To cut costs, several components on the circuit board were replaced with an uncommitted logic array (ULA) – like the video and serial ULAs in the BBC Micro. The chip layout was designed at Acorn using BBC BASIC software written by Sophie, then simulated and verified at the University Computer Lab using new CAD software.

"As the Electron was going to be a volume machine, it had to be 100% right in version 1.0. The order came down, 'Get this debugged now, so it can be ready for Christmas.' Paul was given that job and so, like a re-run of 'The Blues Brothers', he decided to *get the band back together.*

"Paul was contracted through TopExpress to get the Electron tested and ready for production. He recruited Tim Dobson, Kim Spence-Jones, Hugo Tyson, me, and others for an intense period of testing and debugging. We worked some crazy hours and found a lot of bugs in the MOS and one in the ULA. (I almost failed my exams because I was working for Acorn rather than studying.)

"TopExpress had some space in their office above an Italian restaurant in the centre of town. It was like old times at Market Hill – no room to swing a cat – we just got on with the job. Crammed into that room away from Acorn, with prototype Electron hardware, we wrote test cases to validate the operating system. Kim built a circuit board having every expansion interface that might ever be needed for an Electron, just by cutting and pasting BBC designs. It wasn't clever engineering; it didn't need to be. We wrote software drivers for all those interfaces, just to prove it could be done. By the time the Electron shipped, the MOS was about as clean as it possibly could be."

Once more with feeling

"In late 1985, I was called back again to work with Paul on the Acorn Communicator that Christopher had championed. They had firm orders for some, so it needed an operating system – and quickly. At the time, Acorn's engineers were busy with Archimedes and the ARM. There was no one available in-house to work on the project. Again, Paul asked me to write the Econet code.

"It used the WDC 65SC816, which was the 16-bit version of the 6502 – reimagined, bigger, with support for multitasking. There wasn't a direct line between the Communicator and the Arthur operating system, but

Paul Bond and Paul Fellows had lots of conversations about what we were doing, and I believe that may have influenced Arthur's development.

"When I joined, the hardware and MOS were done, so it could boot up but not exploit its extra capabilities. It was to be a 512-kilobyte machine. Software was ROM-based, allowing instant boot-up as a diskless workstation on Econet – ideal for travel agencies and pharmacies. It came with *View*, *ViewSheet*, *VT100 terminal emulation*, BBC BASIC, videotext, and limited cooperative multitasking between modules.

"It was ideal for business terminals in vertical markets such as travel agencies and pharmacies. The operating system was never ported to other Acorn platforms like the BBC Micro or Archimedes. It died with the project and is a fascinating example of Acorn's attempt to build a self-contained business system – lean, networked, and ROM-driven.

"The Communicator OS could multitask, and it did some things in a more abstract way, as we would have liked to have done on the BBC Micro but would have been too much for a 6502 with only 64K of memory; but since we had more headroom, we could do more. Multitasking meant a travel agent could look up and book holidays using the Prestel terminal; switch to the text editor to write up the itinerary and quotation; then open a spreadsheet to calculate costings and commission.

"The Communicator was interesting, but a dead end as Archimedes was coming. Travel agencies bought Communicators, replacing their stand-alone Prestel terminals and separate word processors. It served its purpose but didn't change the world. It was a project that should have been a winner but was overtaken by events beyond Acorn's control."

Spinoff businesses

"Acorn began as a small firm that engaged contractors and moonlighters. It became a proper company, employing its own staff. Some started their own small companies doing contracting work and building their own projects. Some of those ideas took off so they gave up consulting too.

"Most of us who were around back then followed that route. Kim started SJ Research. He had a few product ideas of his own. I brought him Econet for the 380Z from Felsted School. We expected Acorn to take it on – because in the early 1980s, Acorn was doing everything. By the time it was ready, Acorn had become a proper company. The project was too small for them, so Kim picked it up and made it a success. There were lots of other companies around Cambridge that had been involved with Acorn, and who went off and did their own thing in just that way.

"An Acorn web of companies existed where we often worked together; but there was a separate group in Cambridge doing the same thing, spinning off from Sinclair. For me, the two worlds met about 20 years later at Amino Communications. My bosses were Paul Fellows from Acorn and Jim Westwood from Sinclair. Finally, Acorn and Sinclair people were re-united. I think Clive and Christopher would have been pleased."

00001110

Arthur Norman

Dr Arthur Norman is a Fellow of Trinity College, Cambridge, where he distinguished himself in a long career as Director of Studies for Computer Science, and lecturer in the Computer Science department. He could be described as a *force célèbre* in Cambridge circles. He even boasts a fan site due to his influence on generations of computer science graduates. In 2000, Martin Harvey wrote this dedication to Arthur in *Multithreading – The Delphi Way*: 'There is no limit to the eccentricity of university lecturers. He deserves a mention for some of the best quotes ever to fall from a computer science lecturer's lips.'

It's worth mentioning that there are more people on Earth able to perform brain surgery than there are expert developers of high-level language compilers. But don't be concerned; what Arthur will tell you here isn't brain surgery!

When presented with this observation Dr Norman responded by checking World Data to confirm that people developing compilers number about 25,000 while there are about 67,000 brain surgeons on the planet. That attention to detail reflects this man's approach.

In a partnership outside of his university responsibilities, Arthur Norman, Alan Mycroft and John Fitch developed the C language compiler, which was vital to the runaway success of the ARM chip. This is Arthur's story…

"By the second half of the 1970s there was a technology revolution underway with one-chip computer cores (microprocessors) and memory (RAM, ROM, PROM) developing at breakneck speed. To learn more about computers that didn't have to sit in large airconditioned rooms, I attended meetings of the Cambridge Processor Group where I got to know Sophie Wilson and Steve Furber.

"It wasn't long after Acorn was founded that BBC television planned to promote computer literacy through a series of programs linked to a new personal computer. Acorn won the contract to design and build it, and their BBC Microcomputer was a huge success. It felt amazing to see prototypes of it, and to discuss its design with people I already knew but who were now central figures in its development. I even contributed code for BBC BASIC's mathematical functions – such as *log* and *sin*. As a result of my modest contribution, my name was among the list of credits burned into the computer's MOS ROM.

"Later, some of my graphics were used in *Acornsoft Chess* and I co-authored the book, 'Lisp for the BBC Microcomputer'. The academics at the University Computer Laboratory had close links with Acorn and other local companies, such as Sinclair. In fact, some of those we taught were working in those businesses and happy to engage with us.

"As time passed, Acorn needed to consider its next step – beyond the BBC Microcomputer. While excellent in its day, the 6502 processor was being overtaken by newer devices. Computers working in eight bits at a time were being overtaken by machines using 16- and even 32-bit chips. I helped to evaluate the Motorola 6809, which was promoted as a natural successor to the 6502, but we found it didn't provide a big enough performance jump.

"Acorn was considering the NS32016 processor and needed a cross-assembler to convert assembly language programs to machine code on a BBC Micro. With some help, I wrote the cross-assembler. The path I took involved inventing a new high-level language (which came to be known as Arthur's Funny Language, AFL); a scheme that mapped it onto code for the BBC Micro. The Assembler itself was coded in AFL. This showed a few people at Acorn (who may have skipped some of my lectures) that I understood how to implement computer languages.

"A few years later, Acorn designed its own fast microprocessor, the Acorn RISC Machine (ARM), by making its internals as stripped down and simple as possible. Because of my past contacts with Acorn's directors, engineers and computer scientists, I got to see details (under non-disclosure) quite early.

"Computers using ARM processors were expected to have considerably more memory than the older 8-bit machines – by a factor of about 100. Application and system software for earlier machines were commonly built using a machine-native assembler program to convert assembly language instructions (written using mnemonics) into binary machine language programs. That wouldn't be feasible for most ARM users. For them, programming would need a compiled high-level language. A compiler is a suite of software packages that convert from a more-or-less human-readable high-level language into the machine code that the computer can perform.

"The first compilers emerged in the 1950s. By the end of the 1960s,

BCPL was the dominant programming language at the University of Cambridge Computer Laboratory, developed by Martin Richards. BCPL remained important in Cambridge and many of the researchers there understood the language and how it was implemented rather well.

"In the early 1970s, a derivative of BCPL called C was created at Bell Labs; and over the following decade, it grew in global importance; in part because it was used to develop the Unix and later the Linux operating systems. By the time the ARM chip was developed, C had become one of the most important languages for developing complex applications.

"I learnt that Acorn needed a C compiler for their new processor and had been seeking a suitable source. It seems there were serious concerns about high up-front costs and how good the compiled ARM code would be. My colleague Alan Mycroft and I weren't privy to Acorn's internal discussions, but on finding out about the ARM chip, we talked about building a C compiler for our own interests. Being a successor to BCPL, which we both knew inside out, we thought it was achievable.

"It happened that a formal international standard for the C language was being developed. That would mean we had a clear specification for any work we did. We could read and understand the literature on emerging techniques for high performance compilation and see how they would fit with the ARM design. Full of the optimism of youth, we thought we might be able to create a C compiler for the ARM 'over the summer'.

"This was highly speculative, but Acorn lent us a BBC Micro with an ARM-based second processor, at a time when very few were available. It had computing power close to departmental servers of the day, and that could really help speed our work.

"We started with a very limited set of development tools. As we wanted to build our new compiler in C code, we needed an existing C compiler to build it. This is a common problem when bootstrapping a programming language onto a new machine. Happily, *Dr Dobb's Journal* had published full source code for 'Tiny C' a few years earlier. It was only about 1,000 lines of code. A variant of Tiny C called 'RatC' was described in a book by Berry and Meekings that was set up to easily target new or variant computer systems. I planned to use the RatC source to build a simple compiler that could generate a subset of C in ARM code. As I typed in each line of code, I adapted the parts that produced machine instructions to generate ARM machine code. I didn't worry if the code was pretty or compact or fast – it just needed to work.

"Since there was no C compiler for the ARM processor, it couldn't be compiled on an ARM machine. I borrowed a computer that did have a C compiler long enough to get it working. That didn't take long. The result was that I now had a RatC compiler that I could use on real ARM hardware in a state where it could rebuild itself and be used to start compiling other programs, including one that would become our own compiler.

"From there on, Alan and I conducted two activities in parallel. One was to gradually upgrade the version of RatC we were using to make it more complete and more suitable for our needs. The other was to work on our own complier, which was code totally separate from and independent of RatC.

"Since I have been cautioned about discussing brain surgery, I shall skip describing the steps in detail, however it's fair to say that Alan took the lead with reading, understanding and interpreting the emerging ANSI standard for the C language. The draft standard was well advanced, but there were still plenty of ambiguities – where we needed to think hard about what strategies could allow us to meet its requirements.

"Meanwhile I created first drafts of a large part of the new compiler's code. Alan then reviewed its programming style and ensured that details were correct. Sharing responsibilities was quite informal, but we both worked together comfortably and effectively. What we ended up with was a joint effort that neither of us could have completed alone.

"An issue that probably neither Alan nor I appreciated when we started was the scale of the work involved in building a C library. While the core and most heavily used parts of it were easy, the range of features that needed implementation and the level of detail in the ANSI standard meant that we had a mountain to climb. I believe I took the lead on floating point support while Alan introduced some very clever tricks that could speed up some of the string processing functions.

"Once our *Norcroft C compiler* (NORman/myCROFT – or just *ncc*) could build and run small test programs, we showed what it could do to Sophie Wilson. Her feedback on how to improve the code it generated was always useful. A critical point was reached when it was stable enough to build itself, so we no longer needed to rely on the RatC-derived scaffolding. From then on, the way we assessed new versions was by getting the updated compiler to build itself. Our measure of success was when the existing and new versions behaved identically.

"A benchmark for measuring performance at that time was a program known as a *dhrystone*. We were amused to find when we first compiled this test using our new compiler, the code generated was tiny and ran in 'zero' time. We found that the dhrystone test timed the execution of some operations, but its output didn't depend on the results that those computations led to. Our compiler was clever enough to observe this and took the view that if a calculation was going to have its output ignored there would be no point in performing it. We modified the benchmark so everything to be measured was included. It showed *ncc* to be typical of a newer class of compilers that analysed the programs they were working on, giving it a chance to do better than many (but not all!) of its predecessors; and far superior to most other compilers for small computers available at the time.

"As this work progressed Acorn was watching and probably agonizing about whether *ncc* could fill their needs. After all, Alan and I weren't Acorn employees, and no contractual agreements where in place. Buying in services from a proper commercial body with a proven track record of compiler delivery would have seemed a safer route (if less affordable). Eventually, we came to an agreement that satisfied both of us.

"Alan and I continued working on the compiler to get it close enough to meeting the emerging international standard. In that, we were joined by John Fitch. John was at the University of Bath, but as Fitch-Norman, he and I had done business with Acorn on several earlier projects. But now, rather than having all development done by Alan, John and me, Acorn set up the Program Language Group (PLG) to work jointly with us. We three academics founded Codemist Ltd as an entity to represent our side of this arrangement. There were several expectations: Codemist would continue to contribute until the compiler was of reasonable production quality. It would not guarantee indefinite maintenance or support; that was to be done by Acorn in-house. For that to be feasible, the Acorn people needed to understand all the code that existed. For as long as both parties were involved, all project code was available to all of us. Acorn had exclusive rights to use the compiler on ARM systems, while Codemist could use it anywhere else (at the time, other targets for it were hypothetical). This gave Acorn security as it had in-house knowledge and competence to maintain and extend the product. It avoided Codemist finding itself obliged to fix bugs forever; and the amount Acorn had to pay was much less than if it commissioned an external contractor.

"This arrangement could only be acceptable to Acorn once the prototype compiler was close to completion, so they could feel confident they would receive a working compiler. It is well known among software developers that when a project is 90% complete, 80% of the work remains to be done – and that was probably true when the contract between Acorn and Codemist was signed.

"Lee Smith became head of the PLG in 1986 when one of his bosses, Jeff Tansley, asked him to 'help sort out the mess Acorn had got itself into with C compilers'. In other words, he wanted Lee to mastermind the transition from an informal and experimental compiler development with no clear obligations, to a formal joint development. He and Harry Meekings joined in as members of the compiler development team. As well as interacting with Codemist and making substantial technical contributions to the compiler itself, they also worked on the surrounding software infrastructure and shielded Codemist from any distractions. Lee and the PLG made copies of the compiler available to other Acorn contractors and major customers; and in due course, packaged and released the compiler as part of the ARM Development Kit.

"From my perspective within Codemist, I could see little beyond the compiler and the output it generated, along with the C library that users adhering to the ANSI standard could rely on. Through the PLG, Acorn needed to consider its use with RISC OS (for the Archimedes) and with RISC iX – Unix for use in professional and scientific applications.

"In time, the ARM processor was adopted by the embedded market and for use in mobile phones. The original Reduced Instruction Set (RISC) strategy that drove ARM's initial development started with the view that memory was becoming cheaper, so code density was of secondary importance. For phones this was not acceptable, so the ARM chip was given 'thumb' mode, which stored programs in a more compact form. The PLG adapted *ncc* to generate code for that mode, and by achieving that, established its ability to make significant adjustments to the compiler without needing Codemist. Using the same components, they also built Fortran and Pascal compilers. These developments showed that Alan and I got the structure of things close to right at the beginning.

"Eventually, Acorn/ARM and Codemist drifted apart as the compiler became a stable product – no longer an active development project. Codemist adapted the compiler to target IBM mainframes that cost millions of pounds rather than ARM systems that might have been in the thousands. Working with University of Cambridge Computing Services led to the C compiler becoming available across the university.

"We shipped an IBM mainframe compatible system to Japan where it was used on some large computers. We also ported it to a few small and sometimes slightly eccentric machines being built by other UK companies.

"Our unique selling proposition was that we could create an initial version of a compiler very rapidly, and most of the important optimisation was done in machine independent parts of our system. We followed the 'we get an initial version going and train you to take on the maintenance, then we step aside' approach. With that, smaller companies could often get an idea of how things worked early on and obtained a product to offer their own customers at lower cost than they otherwise might have.

"I'm proud *ncc* remained ARM's main compiler and I'm grateful for the continued support among users after all these years; and the company that took over RISC OS when that part of Acorn's business collapsed still views it – after all these years – as respectable.

"It's worth noting that the first release of *gcc* was in 1987. Why didn't Acorn use *gcc*? – the answer is simple: It didn't exist, and it was some time before it gained wide distribution and stability to be useful. By then it was huge and working with it was a much bigger job than working with *ncc*.

"C was not commonly used in Cambridge when we started. Linux was released in 1991 so before then, it was proper Unix, with a mix of licenses, typically only available on expensive workstations. The sorts of computers that Unix ran on were not available to us. We were in a decade where hardware improvement was rapid; but still, an IBM PC or clone was a luxury item and a CP/M-86 second processor sitting by a BBC micro was a lot more affordable. A decent C compiler running on an ARM processor represented a breakthrough in affordable computing.

"The ARM processor has had a huge impact on the development of personal portable devices (organizers and phones). Of course, it was the low power consumption of the ARM chip that made it all possible. But if there had not been a high-quality C compiler for manufacturers to develop software for their portable devices, progress would have stalled. The Norcroft C compiler became the standard shipped by ARM Ltd; so huge numbers of portable devices have incorporated code built using it, developed by legions of software developers.

"The only way I can describe my feeling about that is: WOW!"

00001111

Brian Cockburn

In 1982, Brian Cockburn joined Australia's Acorn and Sinclair distributor. He was one of two technicians hired at the same time, but missed having to deal with the growing pile of broken Sinclair ZX80s and ZX81s.

A few years later, Brian moved to Acorn where he contributed to Econet development. When he arrived in Cambridge, his coworkers believed they already had enough Brians – Brian Jones and Brian Robertson. They were inspired by Monty Python's 'The Bruces', which claims all males in Australia are called 'Bruce'. That's why this is Bruce's story…

"After completing Year 12 in 1976 I spent a gap year working at a Melbourne electronics supply store. I followed that with two years studying engineering and science at Monash University, but formal education was not for me. In 1980, I started working for electronics and gaming machine developers. Working with microprocessors just made me want to learn more.

"Toward the end of 1981, I answered an ad for a technical services role with a small computer supplier in Melbourne. Rob Napier interviewed me. He must have been impressed with my choice of T shirt, which has always been my idea of formal wear; and he didn't seem to mind when I asked for some pallet space in the warehouse. He didn't even ask me why I needed it. He found out soon enough when the Coca Cola delivery truck dropped off a palette of Coke at the loading dock – so started my new career.

"Sinclair ZX machines and Acorn Atoms were of similar build quality, which wasn't very good, so we spent a lot of time making up for their poor design. The Eurocard Systems were a different matter and showed what Acorn could do when they weren't building to a price."

Debugging early releases

"When Acorn started shipping BBC micros to Australia, they were only sending about 50 or 100 at a time. The demand for more machines was always there. It would be a couple of years before we had stock in the warehouse for more than a few days. Software updates were usually once a month, delivered in EPROMs by courier from Cambridge. Disk drives were locally sourced and built by us.

"The Machine Operating System (MOS) started as version 0.1. Storage was just the Cassette Filing System (CFS). The Disk Filing System (DFS) and the Econet Network Filing System (NFS) weren't available when I joined at the beginning of 1982. We received Model As, converted the TV modulator from UHF to VHF, fixed known bugs, and updated them to Model Bs.

"As shipments became more regular, customers seemed to be happier, but we kept the pressure on Joe Dunn and his crew to deliver better working code. They made the effort to support us and sent MOS, DFS, and NFS (Econet) updates as soon as they were ready.

"Most sales were to schools. By the end of 1982, we were regularly shipping Econet networks with System 4 machines as file and printer servers. Orders were coming in from all over Australia, mainly from private schools as they could afford to buy networks. In 1983, I went to Sydney to set up 50 workstations at a Marist Brothers school in one of the most expensive parts of Sydney, overlooking the harbour. That's when I learnt how the other half live!

"Code was stable, though far from perfect, but we had a novel way to approach computer support: We were honest with our customers. We told them what we knew – the good, the bad and the ugly. Even though people had to wait, most were patient and made to feel part of our user community. This was completely different to how Tandy, Apple and our own Sinclair were handled – as a commodity. Acorn was different. We were bridging the gap between retail computer stores and the business systems market. To do that, we wrote books and technical papers, attended education conferences, and ran computer seminars around the country. This approach to treating customers like insiders was new in the education sector. It worked.

"We had different model printers in Australia, so the drivers needed to be re-written. It was my job to test and accept software updates and to modify our printer drivers. It was my first effort at serious assembly language programming. Most of our customers were schools. A small school might only have a machine on a trolley with a TV and printer. Bigger schools had networks of Beebs with file and printer servers. We sold a few for personal use, but that wasn't where our business was focused.

"By the middle of 1983 we had a stable version of NFS. We couldn't get big EPROMs, and we couldn't wait for masked ROMs. We needed a way to deliver an up-to-date MOS, NFS and BASIC. I came up with a hardware hack. After getting approval to fit the update, I spent a few weeks traveling around the East Coast of Australia with a suitcase full of soldering gear and components, updating BBC Micros. Doing technical support in the field was a real baptism of fire.

"Around this time, we received boots-on-the-ground support from Acorn, with a visit from John Coll. He was well known in Australian education circles and was treated like a superstar. His visit did more to lift our credibility than all our hard work. It showed customers we were well supported by Acorn. It also helped them understand why they were paying higher prices in Australia and by then, New Zealand.

"I got to learn how to use a Telex machine, though it's a skill I'm never likely to need again. Later, we were one of the first to have email in Australia. John Coll set up accounts for us with British Telecom. We'd plug a desk phone into an acoustic coupler to call the local telephone exchange. After entering a username and password, it would route us to the Overseas Telecommunications Commission in Sydney. Before satellites, they used undersea cables via the Pacific and Indian Oceans to the UK. Another username and password would connect to *BT Gold*. Then finally, another log-in to access an email account. It was like something out of Monty Python's *Four Yorkshiremen*: 'And you try and tell the young people of today that ... they won't believe you.'

"John Coll's first visit was so successful we asked Acorn if someone else could come out. This time they sent Joe Dunn, head of the Systems Group, which developed the file and printer servers, the MOS, NFS and DFS. Joe and I visited schools in Melbourne, Sydney and Adelaide. People came from as far away as Mackay in North Queensland and from Auckland in New Zealand. Joe presented advanced seminars introducing technical ideas that would help them get more out of their networks. This really boosted user confidence in the Econet.

"After the seminars, systems popped up performing more sophisticated communication tasks – and not just in schools. A large Econet system at the Mackay Mercury Newspaper, and its other papers around Queensland, were typesetting their regional newspapers over Econet on BBC Micros scattered around each office. Rob Napier transmitted the files for his book, *Networking with the BBC Microcomputer*, to the Mackay Mercury for typesetting using this method. This is commonplace now but back then, it was state-of-the-art.

"After Joe's visit, Rob decided that I could be of use in Cambridge. He must have been keen to get rid of me because he didn't waste any time talking to Joe, then shipping me off to Blighty. So, in November 1983, I received an offer from Acorn to come to Cambridge. They offered me a generous salary package, compared to a technician's salary in Australia. I said 'Yep, I like that. This'll be interesting for a 20-something to do.' They really looked after me: I flew across in December, and they flew me back to Australia in February to attend a friend's wedding.

"Cambridge in winter felt unbelievably cold. I'd never known weather like that. Fortunately, the Acorn Silver Building had just been finished, which was a lot warmer than the Waterworks. I worked with a bunch of real nerds – technical people my age, all very clever, interested and dedicated to their work. There was good separation between us in the Silver Building and the management, marketing and admin people in the Waterworks. It never seemed like there was much of a battle between *Us* and *Them*. People from marketing or sales would come out to check progress but usually left us to get on with the job. We did have some supervisors, but they were all technical people themselves, who were happy to let us do what we wanted most of the time. I guess you could call it peer-to-peer management.

"The people doing the technical work all got on well together. It was a culture shock but interesting to go to the pub for lunch with everyone in development five days a week. We had our favourite watering holes. It was all part of the nerd culture that was the norm at Acorn.

"I was a little surprised to find that Acorn really did use Econet to run their business – at least for printing. There were System 4 and 5 file and printer servers, one with a 15"x11" line printer. Still, most people in development had hard disk drives so the file server was mostly used to save masters; and we still used *Sneakernet* (5¼" floppies) to share code.

"I'd found an Econet bug just before I left Australia. They let me get started on that when I arrived, so I was useful from the start. I fixed it!

"Once the DFS and NFS were combined into one chip (DNFS 1.0), there was no free space in the ROM, so every proposal for a bug fix had to be matched with ways to free up any extra space needed. We kept a table that listed how many times a piece of code existed before it was worth making into a subroutine. This was done to a four- or five-byte sequence if the code occurred about six times. We'd look for similar code then work out whether it should be made into a subroutine – all just to save one or two bytes for a new feature or to fix a bug. It was slow, detailed and painstaking work. But fun!

"Everyone writing ROM-based system code faced the same challenges: add features, find the space to make the change, and fix the bug in the code. Once we finished and shipped DNFS 1.2, we had a clean sheet for where we might improve the network filing system for future machines.

"The next NFS development took place in secret. It was hosted on a Model B but intended for the new BBC Master. We had just one wire-wrap Master prototype for testing code. People from Office Systems and Advanced Research and Development worked on the MOS, and we were looking at system design (such as graphics drivers). It was like peak hour on the roads with everyone competing for access.

"We got together over lunch and discussed the design of the Master operating system: what we needed to do, how we'd work with increased space, and the facilities we wanted to provide. Good news: We would have more space for the MOS. To avoid congestion at the prototype, I targeted my code at the Model B. Each time I had a usable release, I'd blow it into an EPROM, and then convince people to test and try to break it in their work machines.

"We were constantly trying to make the operating and filing systems better by adding new functions and features such as more support for numbered networks, larger bridged networks, allowing more files to be open, new buffering to improve the speed of the filing system, and the use of named printers.

"Though it was never sold as a product, we developed an Econet repeater called *The Joint* to extend the length of the network, but this was superseded by Brian Robertson's Econet Bridge – a simple device with two Econet interfaces, a 6502 processor, PROM and RAM. It set up the connection between two networks without needing any user input. It shipped, and I don't believe we ever needed to make any revisions. It stands as an example of Brian's brilliant work.

"We also extended performance of the printer server, adding a hard disk printer spooler. It was written in BBC BASIC by Derek McAuley (later Professor McAuley). Till then, printing in the Silver Building was hit-and-miss. I don't know if it was ever sold but it really improved our lives in that large office space.

"We wrote useful development tools: *Netmon* showed all the packets on the network. It ran day and night, on its own hardware. I don't think I ever turned it off because it was essential for working out what was happening – especially when I was working with the network primitives.

"Another helpful tool was called the *File Server Relay*. It was invaluable for debugging the file server or a station. Nowadays we have professional development tools but back then, we made our own – then we'd go home and make our own beer!

"Before AI, we relied on OI [own intelligence]. Of course, systems were much smaller and simpler then, so roll-your-own diagnostic tools were a lot easier to make.

"Around this time, we started to get Apple LaserWriters, so there was some development of Postscript and Postscript header files to wrap jobs. I was back where I started, doing printer driver support; this time to print spoolers. At that point, we had a great sprawling network. Acorn was spread across three sites: Vision Park in Histon, the Techno Park in Chesterton and Fulbourn Road in Cherry Hinton. I focused on getting networks working together and improving printer and file servers.

"Around this time, customers were finding that their Econet systems weren't as reliable as ours at Acorn. Laurence Hardwick needed help to work with schools and large clients to address reliability problems. With Carl Sellers from Customer Service, I visited some problem sites. The common faults were cheap network cabling and poor workmanship. The Econet, like every cabled network, needs reliable cable, good terminators, and well-built clocks. The people installing the networks were cutting corners. Most networks perform well when installed correctly. What we saw was unacceptable. Considering the amount being paid for these systems, Carl and I were expecting to find professional installations. The worst took several days to sort out. It was disappointing to see that we couldn't ensure the quality products Acorn delivered weren't installed in a way that achieved the best results.

"It was a good lesson to learn. It's not enough to ship a product that works well in the lab; finding a way to deploy it effectively IS part of the product. When I look at 10BASE2, 10BASE-T, 100BASE-T, and even gigabit networks today, I see they really are plug-and-play. It's incredible that this aspect of networking has been solved. People have worked on connectors, protocols and so on, to ensure reliability. Better trained installers, better connectors and cable have all helped."

The Acorn RISC machine

"I worked for Office Systems through Acorn's various reorganisations. Advanced Research and Development was at the far end of the Silver Building from my office. This is where Sophie, Steve, Jon Thackray, and others worked. I didn't get to know them very well, because they were developing products we weren't told about. Most secret was Project A.

"Eventually, I was inducted into the project and asked to make Econet test code to see if we could write the Econet primitives in ARM assembler. I made my test programs and found that we needed one more FIQ Register. Steve Furber said, 'Well, you can't have one extra register because of the way the chip is laid out, I'll have to make it two.' So that's why the ARM's FIQ register set is the size it is. Fortunately, this all happened before the first ARM silicon (the so-called three-micron ARM) was made, which was a raging success.

"Version control is the bane of every software development team. Considering that Acorn's work predates version control systems, it's surprising that code management was never a serious problem. Developers maintained their own source files. Personal releases were passed around on floppy disks. One would be sent down to the drawing office for archiving. The other way we kept track of the code was a printout. When we reached a milestone release there would be an assembler listing printed and bound into a plastic cover. It could be up to 10 centimetres thick. Back then, developers seemed to have an inbuilt discipline. How did we lose it?

"Looking for a bug and to find places where we could save bytes in the assembled code, we'd scan the listing and annotate what we planned to do. So, these huge listings – marked on the spine using a felt pen with something like NFS 3.32, MOS 1.0, or whatever – were the Bible for our development team. Our version control relied on nothing more than that – just programmer discipline, copies on floppy disks, and printed-and-bound source code listings.

"With 8-bit micros, most coding was done by a single programmer, so version control was easier. Once development of software for the ARM began, with multiple programmers on the same section of code, configuration management became a problem. We had several inflatable balloons in the shapes of animals to warn others that a change was about to happen. To make a modification to a particular file I would hang the pink Flamingo above my desk."

Brian Cockburn – on managing Chris Turner

"In case the office was burgled or burnt down, Chris Turner implemented a file storage backup system. Official software releases were stored in a fire-proof safe, but being fully paid-up members of *Anarchists United*, programmers resisted any attempt to organise them. They still took home copies of their work on floppies every few weeks and parked them under their beds. Fortunately, Acorn never had to find out which backup system was more effective because there were no fires and no burglaries."

Chris Turner – on managing Brian Cockburn

"With all the expansion, we quickly ran out of car parking spaces. Memos were sent around asking people to park properly. My office looked down over the car park at the back of the Silver Building. On days when Brian was in a bad mood – usually when someone had messed with his code – he took revenge on the world by parking his car diagonally across four-to-six parking spaces. Those of us managing research and development in the Silver Building would have to draw lots to see who was going to talk to him; sharpening our diplomatic skills, it was great preparation for joining the Foreign Service."

00010000

Carl Dellar

Carl Dellar was born and raised in the Cambridge area. He studied computer science at the University of Manchester where he learnt to play darts and drink Boddingtons Bitter. After being awarded a First in computer science in 1976, he started a paid job writing COBOL. It didn't take long to realize there was more to computing than what he was doing. This is Carl's story…

"I applied to study for a PhD under two giants of computer science, Professors Maurice Wilkes and Roger Needham at Cambridge, and was thrilled when they invited me to meet them at the Computer Laboratory. I knew Cambridge well but had never seen it so congested. Finding a car park was as difficult as getting past road closures to get there.

"Arriving late for my appointment, I was met by a little man who appeared quite unremarkable. I thought, 'This is a posh place. They even have a lift operator!' He said, 'Mister Dellar?' I nodded. As we got into the lift, I said. 'Gosh, I'm half an hour late! The traffic was terrible. I have an appointment with Professor Wilkes.' Then this little fellow said, 'I'm Wilkes. I'm Wilkes.' My shyness and reluctance to say the first thing that popped into my head had saved me again.

"Wilkes and Needham conducted the interview and decided to offer me a place. I was told afterward that I had asked the magic question that got me in: 'Do you build things here? I understand a lot better when I do it myself.' Luckily for me, Cambridge liked practical-minded people. They liked to build things too.

"Just before going to Cambridge, I met a young woman called Debbie. She is the reason I was able to complete my PhD and everything I've achieved since. Debbie was born in Hong Kong and came to England in her teens. She was working for a subsidiary of British Airways when we met. Eventually she came to live with me in Cambridge and worked at the university. I had a small grant, but when that ran out, she supported both of us. I couldn't have done it without her.

"Capability machines were an area of advanced research in the 1970s. They provide good memory protection, which has become a big issue with computer security. I worked on the CAP operating system, rewriting the file and virtual memory systems to use the Cambridge File Server (designed by Jeremy Dion) as the backing store.

"There was a tradition in the Computer Lab – probably still is. Tea breaks were at 11:00AM and 3:30PM. The tearoom is where I met some interesting people and got to mull over whatever problems they were working on at the time. Pamela Raspe, an anthropology PhD candidate, and her friend Allison Smith, a biochemist, used to join us in the Computer Lab tearoom. In 1979 people started talking about Pamela's physicist boyfriend, Hermann Hauser, who had a startup business with Andy Hopper. They produced single board computers that the university used for teaching assembly language programming.

"In early 1980, I was struggling to write my PhD thesis. One day I was in the tearoom, feeling quite glum, when Professor Wilkes asked how I was doing. I muttered some half-hearted reply. Seeing my struggle, he said, 'Can I give you some advice Carl? I've written a few books myself, and I've found the best way is not to start at the beginning or at the end. Start with what you know most about. Write about that first; and whatever you do, write every day, because even if you cross it out later, you'll have gone through that process. It becomes a habit and eventually, you'll get there.' Thank you, Maurice – brilliant advice that I've never forgotten.

"Roger Needham approached me one day with an extraordinary offer – a job in the Computer Lab. Me? A lad from Kneesworth, getting a job offer to work at one of the most prestigious universities in the world! I didn't know what to say. I went home and told Debbie. Her response was unreservedly clear: 'You want me to stay in this cold place for one day longer than I have to?' I had to gratefully decline Roger's generous offer.

"During the summer break, I flew to the USA and Canada to attend job interviews. Debbie and I were engaged to be married. We decided to escape Britain's climate and felt there would be more interesting job opportunities in North America. I flew to the East Coast where I was interviewed for a position at Prime Computer in Framingham, and another at UMass. I went to Canada to be interviewed for a job in Montreal. Finally, I flew down to Texas Instruments in Austin.

"Intel was building the 432 processor – another capability machine. Texas Instruments wanted to build one too (despite Maurice Wilkes' advice that they weren't suitable). He felt they were good for memory protection but too complicated and too slow. The TI team thought they had ways to mitigate this and believed that support for networking would help differentiate TI's machine from what they believed Intel might be doing, so they made up their minds to push ahead. I was offered a position in their development team.

"Late in the summer of 1980, I submitted my PhD thesis. I would be working in Austin, but it was taking ages to get a visa. By then, Andy and I had become good friends. When I told him I needed a job, he said, 'Go and see Hermann; he's got plenty of money.' So, on the day I handed in my thesis, I walked across the road to talk to him. Hermann explained that he wanted to build a file server and promised to think about giving me the job.

"I started work at 4A Market Hill in September. My parents ran a small grocery store in Kneesworth – now sporting the unlikely appellation: Bassingbourn-cum-Kneesworth. Christopher Curry's parents would sometimes shop there. One day, Mrs Curry asked my mother how I was doing. She said, 'Carl has just finished his studies. He's started his first job at Acorn Computers in Cambridge.' To which Mrs Curry replied, 'Oh, my boy Christopher works there too.'

"That winter was unbelievably cold. It was a race to get in first each morning – to get a desk close to the radiator. We built the Econet file server quite quickly. I later wrote a paper published in *Software - Practice and Experience*, November 1982. It explains what Paul Bond, Joe Dunn and I managed to do: *A file server for a network of low-cost personal microcomputers*. It can be found in the Wiley Online Library.

"Laurence Hardwick got the Econet hardware working. It was two twisted pairs; one carried the clock; the other carried the data. The clock speed depended on the length of the cable, usually running at about 250kHz. The Motorola MC68B54 controller chip encoded the data, with help from Laurence's clever collision arbitration. Paul Bond was developing the low-level primitives when I started.

"As I was busy with the file server, I needed someone to get the client software working. Hermann hired Joe Dunn to do that. It started to work very soon after he joined us. His command line processor had incredibly tight coding because the Atom had so little memory. There wasn't enough memory to execute a command like *create file* or *delete file*, so Joe did most of the work on the file server.

"When a user enters a network command on the Atom, it is sent to the file server, where it is parsed, decoded and sent back as a procedure number with arguments. The Atom executes the numbered procedure, using the arguments passed to it. It was a brilliant solution. Joe can be far too modest about his abilities. He came up with all sorts of clever solutions to difficult problems. If the Atom never had Econet, our focus on networking might have died before the BBC Micro came along.

"My Level 1 Econet file server was a compromise. Affordable disk drives stored only 100 kilobytes, and there were time constraints to get the job done. Acorn's approach made a lot of sense: Get a product working and ship it. Let users discover how it fits their needs – what they like and what they don't like – then make it better.

"The file server was extended several times and became an inexpensive and effective solution. The way it read in the whole directory in one block was a poor decision, and the length of names was too short. But those who came after me did a great job with the Advanced Network Filing System (ANFS).

"Sophie's doorstop-sized sandwiches used to make me laugh – the biggest I've ever seen. She'd wander around the offices biting into these huge chunks of bread. We got the file server and Atom working, then on one visit, *doorstop* in hand, Sophie said she needed a printer server, so without further ado, or a design specification, we built that too.

"Some time after Easter 1981 – the file server had been working for a while, so we unveiled Econet in London. Hermann introduced it, putting on a great show. Driving back to Cambridge, Joe wanted to stop and pick up his saxophone. I waited in the car while he went up a flight of stairs to get it. Next thing, I heard a saxophone being played. It was Joe. He was a brilliant musician – so incredibly talented.

"The time was fast approaching when I had to leave for my new job at Texas Instruments. Everything seems a blur after I finished my work on the file server. In June, Debbie and I were married with Andy Hopper as my Best Man. We went to Hong Kong for our honeymoon. On returning to the UK, we packed up all our earthly possessions, then emigrated to the USA early in August. We flew to Austin where I worked on the capability machine – then TI's senior management cancelled the project. In January 1983, we moved to Palo Alto.

"I started developing a database at the Hewlett Packard Labs. People were talking about the RISC machine they were building at the back of the lab. I was just an observer, having nothing to do with it, but stunned to see how quickly it came together. HP's *Precision Architecture* is the result of what was known inside the company as the Spectrum program. Their goal was to move all their non-PC compatible machines to a single RISC CPU family. Planning started a year earlier, defining the instruction set and virtual memory system. Construction started just after I arrived and was delivered to software developers the following year.

"While this was going on, I read a paper by Professor Forest Baskett, a heavy hitter in the world of computing. He worked at Xerox PARC, then joined Stanford's Computer Science department in 1971. On sabbatical, he built the DEMOS operating system for Seymour Cray in 1979. He topped that by supervising four PhD students who have all become major contributors in computer systems design. Professor Baskett left Stanford in 1982 to start DEC's Western Research Lab.

"Compilers such as C and C++ convert high level instructions into native machine code for the computer to execute. In complex instruction set computers (CISC) there are so many different instructions to choose from. Professor Baskett and his team at Stanford studied which machine language instructions were generated when compiling C source code. They found compilers used a small subset of the available CISC instruction set – like load, store, add, subtract, shift, rotate and simple jumps. By using only simple instructions, they realized that the instruction pipeline could be made simple too, allowing the processor to run much faster. I also read a paper about the RISC-based IBM 801 machine, written by John Cocke and others at IBM.

"Sophie and Steve wanted to build their own processor – bigger registers, faster, cheaper. The 6502 has a simple instruction set with a single 8-bit accumulator for performing calculations. It has features like those found in RISC processors. A 32-bit processor built like a 6502 would run much faster than 16-bit and 32-bit CISC processors of that time.

"Early in 1983, Hermann and Andy came out to California. We went to dinner at a Chinese restaurant called *Chef Chu's* in Mountain View. I remember Andy saying, 'We're thinking about building a RISC processor with a wider bus and a simple instruction set like the 6502.' I gave them copies of the papers about the IBM 801 computer and the analysis of the compiler-generated code by Baskett and his students. This contributed to some lively discussions about RISC computers and the benefits of a smaller instruction set. It helped to seal the direction that Acorn would take."

Note. Chapter 00100000 – *Before the Revolution* gives more details about the IBM 801.

"Back in Cambridge for Christmas 1983, I visited the silver building in Cherry Hinton. They had been working on the new chip. Their small group was making great progress. Steve and Sophie oozed confidence. They knew where they were going and how to get there."

How the Acorn Research Center started

"In 1980 – the last year of my PhD studies – Jim Mitchell came to Cambridge on sabbatical from the Xerox Palo Alto Research Center (PARC). Debbie and I came to know Jim and his wife, Judy, quite well. Talking to Jim about Xerox PARC, I discovered a big wide world beyond Cambridge that I knew nothing about.

"When Debbie and I moved to the San Francisco Bay Area, Jim invited me to visit Xerox PARC. Its bleeding-edge development was amazing! I saw the Alto personal computer (inspiration for the Apple Macintosh and Microsoft Windows), a document description language that would drive laser printers, printer servers, and a system with window-type applications running over the Ethernet. Many of those ideas would take years to be released, in products from Apple, Microsoft, Adobe, and others. For example, Robert Metcalfe and his team had invented Ethernet a decade earlier. It was about to be released as an IEEE standard.

"Andy contacted me again in 1984. 'We're good at designing hardware but we would like to improve the way we develop software. Silicon Valley has some brilliant software people. Maybe we can do something in America with some of them.'

"It happened that a lot of people were leaving Xerox PARC. Bob Taylor led it through the golden years of Ethernet development. He accepted an approach from DEC to set up their Systems Research Center. That started a chain of resignations. Adobe was starting up then too, though it was still below the radar. They also benefited from the unrest.

"I told Andy, 'Some of PARC's best people are leaving. Butler Lampson has gone, and Jim Mitchell is looking around. If you are serious, there'll never be a better time to recruit the talent you need. Contact Jim Mitchell; he'd be a great leader.' With that, Hermann called Jim and set up a meeting in Palo Alto. I was invited too. That's how the Acorn Research Center (ARC) started. Jim Mitchell would be running it, and I'd be his left-hand man.

"We were thinking about what to do first. Hermann often spoke about Philips write-once, read-many (WORM) optical discs, which were new back then. Optical disk storage had far greater capacity than early hard drives. They had the potential to be useful for day-to-day operation, as well as for archival purposes, offering much cheaper storage compared to a hard drive. That is no longer true, but back then it cost a small fortune for just 10 megabytes of hard disk storage. I'd heard about optical disks when I was at HP Labs and had been thinking about how to build a file system using them. I wrote a design document describing how to do it and discussed it with Hermann and Jim. They liked the idea.

"We moved to the corner of Page Mill Road and El Camino in Palo Alto, on the 10th floor of one of the towers. As well as my optical disk project, we started to build an advanced operating system for the ARM chip. ARX would have all the best features of a modern operating system with preemptive multitasking and multithreading. It would support an optical disk file system, windowing and an advanced text editor. The OS had to fit into 512 kilobytes of ROM. That size seems laughably small now.

"We hired some brilliant people who wrote ARX using the Modula 2+ language. At that time, we had two of the best compiler experts in the world: Mick Jordan, who had done his PhD at Cambridge, and Trevor Morris. Our focus to begin with was getting all the functions right. It was slow but overcoming the speed problems would have to wait. Adding the SWP instruction to the ARM2 processor helped but strains on Acorn's financial position meant we were running out of time.

"It takes resources to get complex systems to their optimum. Acorn didn't have enough time or resources. We were doing what we believed were the right things, but ARX never made it. It took too long, was too slow and too big. Quite simply, it was too ambitious – a bridge too far.

"Nine months before the launch of the Archimedes, when it was clear that ARX would not be ready in time, Sophie decided to build a less ambitious system, *Arthur*. It first shipped in 1987 as a 'stop-gap measure'. Arthur evolved to become *RISC OS*, which proved to be a very credible operating system, well suited to Archimedes desktop applications.

"I was there in 1988 when Olivetti took over Acorn and the Acorn Research Center. The people leading the Olivetti Lab at Menlo Park wanted to merge the two enterprises. This caused a lot of friction.

"Maurice Wilkes gave me some more advice: 'Life is like a game of chess. It's not the last move that matters; it's the next move.'

"In 1990, Olivetti needed to tighten its belt, so it closed both its labs in California. I left and joined Oracle."

00010001

Colin Priestley

Colin Priestley managed the production of Acorn Systems and Atoms during its critical growth phase between 1979 and 1982. His abilities as a logistics manager helped to keep the company alive during its startup phase. Later, he demonstrated those same skills on the other side of the world when he joined Barson Computers. As Colin explains…

"After reading a review in a computer magazine early in 1979, I popped into Acorn's office in Cambridge to buy a System 1 – so began three of the most enjoyable years of my working life.

"Market Hill was not what you would expect from a computer store. In fact, if you saw a retail outlet like that today, you wouldn't stop. A dark, drab and dingy passage opened directly off the Market Square. A heavy door led to a storeroom. I called out, but no one answered. I wondered if I'd come to the right place.

"Eventually, I found Peter, the only one there. 'Everyone's out,' he said. 'Can I help?'

"I told him I wanted to buy a System 1. He replied, 'I hope you don't want one assembled and tested. We just can't get 'em.'

"I offered to take five home to assemble and return. 'Above my pay grade,' he said. 'Come in tomorrow. I'll tell Chris you're coming. Just ask for Chris.'

"I went in the next day and met Chris Turner and Hermann Hauser. I repeated my offer. Without much discussion, they asked me to join them on a full-time basis. I don't remember what they asked me, but I must have said something to convince them that I could be of use.

"Chris was their Chief Engineer. We worked well together and were able to get a lot of projects moving. He was very talented, but up to his eyeballs in work. He didn't have time to build and test prototypes, so he passed his designs to me, to mock up. I became handy with a Vero pen and occasionally, wire wrap.

"When they decided to fit an Acorn Microcomputer, a back plane, and some Eurocards into a Vero case it became the System 2. This was often shipped with a keyboard, video display board, a cassette-based operating system and Acorn BASIC interpreter. Sophie had written this early version of BASIC while still at university. It would later shape BBC BASIC. Add a 100k floppy disk drive and more memory to make a System 3. A larger case, more memory, and two double-sided floppies made the System 4. The System 5 came much later, when the faster, 2MHz 6502 processor was released.

"I kept busy assembling System 2s to meet the backlog of orders. Space was at a premium, so I shared a bench with an ACE electronic poker machine, which would occasionally burst into life, as new code was tested. In my spare time, I helped Chris as he set up to produce the Atom – checking circuit layouts and parts lists. I helped with a couple of design fixes for it. Spotting the expansion bus connector being the wrong way around (avoiding a full production run of 1000 unusable boards) justified my pay that week.

"There is a well-known photo of what was claimed to be the first Atom being sneaked out Acorn's back door to go to a show in Belgium. Well, it wasn't. The picture was taken by us as a publicity stunt that appeared in *PC World* as a paid advertorial. It was a Eurocard System keyboard case, with a sheet of black card stuck behind the front opening and a full set of key tops glued to the card. Luckily, the show stand had a Perspex screen that prevented too much scrutiny.

"The advertising sign boasted a 6502 CPU with 24k of memory, a user-definable boundary for graphics and program use, and a full keyboard, which sold as a kit or fully assembled – *at a price you can afford.*

"Even without seeing a working demonstration, people were sold – and so were hundreds of Atoms. I was overwhelmed with orders to fulfil, so Market Hill quickly expanded into a store, pack-&-dispatch, and repair area. I recruited about a dozen new people. We soon ran out of space, so my crew all moved to Milton Road; while the development team, including Sophie Wilson, Laurence Hardwick, Chris Turner, Hermann and Christopher stayed at Market Hill.

"From the beginning, lack of space was a problem. As production, shipping and maintenance crews expanded, the problem became critical. Even after the marketing and administration team moved to offices in Bridge Road, there was no relief. Some of us moved to spare space at French's Mill, owned by our System 1 and Atom assembly contractor, Ray Fordham. We stayed there while old university premises at Milton Road were refurbished. Ray continued to support System and Atom manufacturing for another six months after we moved to Milton Road.

"At that time, Acorn's finances were always tight. But with a little care and planning we were able to weather the storm. There were several weeks when we literally had no money. In normal business you have creditors and debtors. With Acorn at that time, I had to keep creditors on side; and the same with debtors, who appeared as a negative in the balance sheet, as they'd paid for their Atoms and were waiting for delivery. Our problem was: we couldn't get cases fast enough from Wong's in Hong Kong. We fielded hundreds of calls from customers demanding refunds. Most people were understanding when we explained the hold up. We shipped the kit without a case to most of them, so they could get on with assembly, and the case followed later.

"One weekend our people were rushing to get ready for a tradeshow at Olympia. It was different from most shows we'd attended as we could sell product from the stand. We were short of machines to sell but had just launched new games for the Atom. We hoped selling software would generate some much-needed cash. I hadn't intended to be at the show but got a call to send more games. The crew was flat-out shrink-wrapping software. I was expendable, so I drove the overloaded van carrying the packages. After delivering the load to a grateful salesforce, I spent the rest of the day looking at our competitors.

"While standing on the apron of our stand with a coffee in my hand, a visitor asked, 'Do you work for Acorn?' I had failed to stay invisible. 'Ah, well, yes, I do, but I'm not here today in an official capacity. You need to chat to one of the sales team. I'm production manager not a salesperson.'

"Ignoring my attempt to disengage, he chatted for a while. Finally, he asked me why he should buy an Acorn, rather than a Sinclair or Commodore. I have a well-deserved reputation for being quite acerbic. I think I said: 'Why would I try to convince you? We have more orders than we can cope with now, so adding you to the list will just make my job harder.' He laughed and left it there.

"On the Monday morning, I was called to Christopher and Herman's office. The chap from the stand was there too. I thought I was in trouble for making that comment but to my surprise, he wanted to place a large order for Atoms to become part of a business package he was marketing. But a condition of the deal was that I was to personally look after him. How perverse! All I had done was say I didn't want his order.

"On most days, the inwards mail gave us a laugh. Shipping so much product, we expected to get returns, but far more than expected. Most were correctly packaged and had obvious faults, the most common being electrolytic capacitors fitted the wrong way (even though CBT's hardware manual stressed how important it was to fit them correctly). He explained they were polarised and how to work out which end was which. With thousands sold, some still got it wrong.

"We received a faulty Atom back with a lovely letter from a Royal Air Force engineer, who explained how much skill he had and how much care he had put into the kit's assembly. He had checked everything thoroughly but simply couldn't understand why it didn't work. He suggested: 'It must be a board or component fault.' It was so neat, just as he described it: 'perfect'. We looked at the board and turning it over, the penny dropped! The components hadn't been soldered to the board; they had all been glued in place. I felt sorry for him and wanted to acknowledge his effort. I decided to send him the best-looking homeless orphan we had and gave him a free upgrade to a full spec machine. We kept his machine as a souvenir.

"A lot came back correctly packaged but with nothing explaining the fault or more importantly, who had sent it. At one point we had 17 Atoms with nothing to identify the owner; but worse than the problem of the Atoms with no owners, was the pile of empty boxes that were returned with no customer details and no contents in the package. Nothing. The postman was delivering squashed-flat empty boxes. At one point, we had at least a dozen of them. Outbound kits? Assembled kits for us to fix? Full specification, assembled machines coming in for a refund? No idea… And oh joy, the inevitable phone call: 'Did you get my return?'

"One day, our postman came with two Royal Mail 20 litre plastic bags full of Atom keyboards, main circuit board, loose components, and a few manuals. But the empty boxes gave me a clue: It looked like they had swept up whatever had fallen out and just placed it all in a bag for us.

"My biggest regret was a letter I received from a lady attached to an Atom kit returned for refund. She wrote that her husband had cancer, so he thought he would like to build an Atom to play with. Sadly, we had taken so long to supply the kit, he passed away before receiving it.

"One cold and miserable Sunday morning in late December 1981, Christopher Curry called. 'I need a favour. Can you bring your van and meet me on the old wartime airstrip at Bourn?' That sounded mysterious, but interesting. 'Yeah sure. On my way.'

"Bourn is located about seven kilometres west of Cambridge. In 1941, it was a typical wartime bomber airfield. Most of the buildings and the control tower had gone but the runways were still there.

"When I arrived, there was nothing and no one else around. Finally, a helicopter flew over, landed, and shut off its engine. 'I barely have enough juice to start her up again,' the pilot said. 'I thought I was coming to an active air strip.'

'What are we doing?' I asked.

'I have these boxes for you.' he replied, as he started to unload about twenty BBC machines.

"The machines had been flown from Cleartone's plant in Wales. It was snowed in, and their first batch was ready to ship. Christopher arrived and asked whether the press had been. The whole exercise was a publicity stunt to show how much we care: 'Bugger the expense, we must ship the product!' was his imagined headline. It's a shame no one thought to contact the media.

"We were a close-knit bunch and worked well together; but I was ready for a change. One day as I sat at my desk, the phone rang. It was Rob Napier, Brand Manager for Acorn at Barson Computers in Australia. He told me he needed a capable production manager and offered me the job. I got him to hold while I phoned my wife on the other line. She had just been offered a redundancy package. I asked if she wanted to move to Australia. 'Yes,' came the reply.

"Christopher had given Rob the okay to poach me, but once it was known I was leaving, I was asked to reconsider. The University of Cambridge and Control Universal offered management roles too, but Australia seemed like a chance to broaden our horizons, and here we are to this day."

00010010

David Bell

1982 was a challenging year for the British Broadcasting Corporation and Acorn Computers. The first BBC Micros shipped in February but demand outstripped supply by a huge margin. Being sponsored by the BBC, consumers expected everything to run without a hitch. Acorn was a small startup working in cramped offices around Cambridge, but they were facing a Tsunami. Despite their best efforts, there were production delays, quality control issues, poor customer service, weak technical support, development delays, logistical problems… The list goes on.

To save its reputation and the project, the BBC decided to bring in an external consultant. One of its recommendations was for Acorn to employ an experienced middle-level manager to liaise with BBC Enterprises and other departments that contributed to the *BBC Computer Literacy Project.* The person recruited to fill that role was David Bell.

David remained at Acorn throughout its highs and lows, finally leading the team that developed the Archimedes series. Here is David's story…

"It was January 1983. The BBC Micro had been released a year earlier, and Acorn was drowning in its own success. I had recently returned to the UK after working at Xerox in the USA. Prior to that, I'd been with Marconi Instruments in engineering and support roles. I applied for the position of BBC Project Controller in response to a newspaper ad. The recruitment process was unusual. It was the only time in my career that I was interviewed by both the employer and the customer.

"The role was described as liaising with BBC Enterprises (though other departments would sometimes need assistance). I was recruited by Acorn, and interviewed by Christopher and Hermann, but the appointment also had to be approved by John Harrison of BBC Enterprises. Being the link between Acorn and the BBC, both parties needed to be happy.

"For the first year or two, I trod a deep path between various BBC departments and the design, production and delivery teams at Acorn Computers. 'Project Controller' was an empty job title. I had no special power or authority to control anything. What I did manage to do was build clearer and more responsive lines of communication between various departments of one of the world's largest broadcasters and a small, but rapidly expanding group of brilliant people in Cambridge. To be fair to everyone involved, they were being asked to do something incredibly ambitious.

"David Kitson, Head of Transmission Group, Designs Department, was effectively the BBC's technical director. I dealt with him on major issues and two of his staff, Richard Russell and Jim Day, on day-to-day matters.

"Liaising with the BBC about product and development timelines, my life was full. I was the first point of call when they received complaints that couldn't be resolved. I also attended meetings with BBC Enterprises to review overseas sales. Through that connection I often met with Bob Bayham from Acorn International. Christopher Curry worked with Bob when he managed Sinclair International and helped Hermann and Christopher establish Acorn in overseas markets, while still working with Clive Sinclair. Sadly, Bob died while I was there at Acorn.

"We'd have weekly progress meetings in Cambridge with John Radcliffe, Jean Nunne (BBC Education), David Kitson, Jim Day and occasionally Richard Russell, Roy Williams (BBC Merchandising) or David Atherton (BBC Publications). Christopher or Hermann would sometimes attend to address some of the more awkward questions. When I wasn't in meetings, there were never-ending telephone calls. A BBC hotline was installed in my office so they could bypass the busy switchboard. Being available to them was always a priority.

"No one inside Acorn had a specific commercial role when they won the BBC contract. I arrived after the royalty deal and performance commitments were already in place. Customer Service was knee-deep in complaint letters and phone calls. It was an unmitigated disaster. Very little was being delivered and there were serious quality problems as production ramped up.

"Acorn had a tiger by the tail, but Hermann and Christopher were astute enough to know they couldn't deliver what the BBC wanted without recruiting some experienced senior people such as Jim Merriman, the Manufacturing Director. I was one of the middle-level managers, recruited about the same time as Mike Bicknell (Customer Support), Phil Smith (Manufacturing), Barbara Cole (Purchasing), and one or two others. We were there to address what was missing at Acorn: people from large commercial operations who could help to sort things out. Phil came from an ICL spin-off. He brought strong manufacturing experience – a really, good guy who worked for Jim Merriman.

"I had a technical background as an electronics engineer. I joined Acorn when I was about 38 years old. Even with my experience at Marconi and Xerox, there was no way to prepare for Jim Merriman – a force of nature – this fiery Irishman is a nice guy, if he accepts you. He did give me a rough time for a week or two after I joined; then things settled down.

"Prior to Mike Bicknell joining Acorn, there were some well-intentioned and very bright guys in technical support doing their best, but deliveries were up to a year late. I worked closely with him while he sorted out Acorn's response. Considering what he had to overcome, he did a fine job. An excellent manager, he set procedures in place and brought his staff together as a team. Despite his hard work, Acorn was delivering hardly anything of note, so the complaints continued and people at the BBC were tearing their hair out. It took time to fix the chaos that had been created.

"Acorn in the early eighties was a young, rapidly growing company. No one knew exactly what would happen next – or what to do next. Everyone was firefighting, doing their best. There was no company structure. It was almost dead flat. Not surprisingly, it appeared strange to the BBC.

"Christopher's focus was marketing. He loved to build relationships. He would go down to London and meet people, while ignoring most of what was going on in Cambridge. He's very bright, very funny, but he really didn't have much hands-on input into the commercial side of the business. Though I worked for him, he took little interest in what I did.

"Hermann is also very bright, focused on research and blue-sky ideas. He didn't spend much time with me but was always available if needed. He focused on what was happening outside Acorn and on research and development. They both left most commercial decisions to the senior management team – Jim and the others who were running the show. That's not a criticism, though there were times like the Electron and US/Canada BBC Micro disasters where a tighter rein might have made the difference.

"Christopher and Hermann didn't understand how long it would take to ramp up the supply chain and service delivery. The dates they promised the BBC could never be achieved. It wasn't deliberate; they just didn't appreciate how much time it takes to get things into manufacturing; and to get the volume up. Other than Wong's in Hong Kong, they had little experience working with delivery companies. It was just a question of getting up to speed. The problems faced by the BBC and Acorn when they failed to ramp up production were due to Hermann's and Christopher's unbridled optimism, their inexperienced engineering team, and the failure of the BBC to understand the challenges when it made unrealistic promises to the UK government and the people of Britain.

"What no one managed to do effectively was deal with supply and demand. The problems didn't end with manufacturing. There was continual conflict at Vector Marketing's warehouse because they were dealing with orders sold directly to individuals, through dealers, and through distributors. Working through different channels was confusing. Later there was the US/Canada version. While they weren't big numbers, it added to that confusion. It took a few years for Acorn to master its supply channels.

"Acorn had grown by finding brilliant young graduates, providing minimal support, planting them in uncomfortable and cramped working conditions, feeding them large servings of free pizza (with rice and chips), then standing back and waiting to see what happened.

"The BBC was founded on rational management traditions. It needed to learn the status of work at regular and well-run planning sessions, all designed to ensure that Acorn delivered what it promised on time and on budget. Somehow, without any special powers but an organised mind, it was my job to work some magic. With so many delays in product development, there was an urgent need for someone to keep track of the various 'Cheese Wedge' projects – like the Second Processors, and the Teletext and Prestel expansion units; so that's where I started. I later picked up planning and liaising with Acorn's office in Massachusetts for the USA/Canada version of the BBC Model B.

"When the need for close liaison with the BBC passed, my next role was as Technical Support Manager. I moved with the Customer Services team and other technical staff into the new silver building in 1984 as the need for support from third party software and hardware developers grew. I project managed the BBC Master's product development (Project B) and with the team, produced first class reference manuals and user guides. They performed brilliantly, leading to what was a successful Master 128 launch. Acorn was coming of age as a hi-tech business.

"Project A had been a closely-held company secret, known only to those who had been briefed in – a process more commonly seen in military projects. It turned out to be Sophie's and Steve's ARM chip (the Acorn RISC Machine), developed in parallel with Hermann's Business Computer. In 1985, our group moved to Newmarket Road, leaving space for the ARM team to grow. Acorn had split into divisions. Later that year, my new boss, Jeff Tansley, asked me to take over Project A. The original team was being disbanded; the ARM was no longer a secret; and the next phase was to develop the Archimedes. I guess management felt that 'Project A' was such a clever name, they should recycle it!

"I was asked to commit to a June 1987 launch for a project I knew nothing about. It was to have been a business machine with the ARX operating system but now needed to be both a low-cost BBC Master replacement (Archimedes 305/310) and an Acorn business variant (Archimedes 405/410). One Thursday morning, I went into the recently-vacated project office and rifled through the abandoned paperwork to find out what I was now managing – interesting times.

"I was able to pull together my trusted Project B team for this assignment and Jim Merriman promised that I had the full support of his manufacturing team for this tight schedule. We moved to new offices at Vision Park, and I was given the grand title of Support Group Manager, responsible to Malcolm Bird. I was also supporting customers, third-party developers, and of course, the BBC. In my spare time, I led development of the MIDI module and the SCSI Podule.

"In June 1987, I was again filled with pride after attending another successful launch. The team had taken an unknown, half-developed product that needed radical changes on a very tight schedule. Despite the hard work of everyone in the operating system team, the Arthur OS was not the final goal for the Archimedes, but it did meet their objectives. That would come later with RISC OS. One small win was publishing two Master reference manuals. We managed to pull together new features and present them in digestible form to an audience that still praises those publications. Documentation is often seen as a product cost. In this case, the manuals added several hundred thousand pounds to Acorn's bottom line – well worth the effort.

"In the early 1990s, Malcom Bird dropped by my office to tell me he was moving to a new role in a new department where he would be managing the new Acorn Network computer with Oracle. He wanted me to take over his role as Technical Director (Acting). I would be managing the 73 engineers in the development team. I was still *acting* in that role in 1995 when, along with 56 others, I left the company as Acorn's time was coming to an end. It was time for Arm to take the technology forward – which it did in spectacular fashion."

...

Reflections

"When Acorn started, it was made up of very bright people. I've worked at other companies with some bright people – Xerox, for example. Acorn Computers is what happens when there is a concentration of really, clever people who happen to be in the right place at the right time and have the courage to take advantage of opportunities.

"They came together almost by accident. They weren't there because of their business ability, so lots of things did go wrong. Everything relied on individual motivation and a personal sense of responsibility. No one tried to control anyone. It was a very open environment. You were expected to see what needed to be done and get it done. Most of the time that worked spectacularly. When it failed, it was equally spectacular."

00010011

Jes Wills

Jeremy Wills, better known as Jes*, is often mentioned in conversations about Acorn. A highly capable computer scientist, he extended Econet performance and developed verification and validation software used to test the first generation of ARM chips. Now living in Austin, Texas, this is how he recalls his time with Acorn…

"My university studies at Cambridge followed the same path as Sophie's, and a few other Acorn alumni. I started with a year of mathematics and followed that with two years of computer science – a course that was in its infancy back then. As graduation loomed closer, I started to think about getting a job. In May 1981, a message was posted on the noticeboard – signed 'Professor Hopper'. Andy was looking for people to work at Acorn Computers.

"It was a Thursday when I went to see him. I remember our brief conversation very clearly. You could hardly call it an interview. He just asked, 'When can you start, Monday?' I thought I better not seem too keen. Exercising my worldly experience, I replied, 'I can't start till a week on Monday.' Andy seemed happy with that. He said, 'That would be great. The office is behind the Eastern Electricity showroom in Market Square. Just show up, say who you are, and they'll be expecting you.' That how I started. We didn't discuss money. We didn't discuss any details, really. I had no idea what I'd be doing. That's how things happened back then.

"When I showed up, people were heads-down, staring into the embryonic wire-wrapped Proton. I started on the Econet. The BBC Micro's specifications were still evolving. Steve and Sophie threw out ideas, trying to convince the BBC that such-and-such a thing was a good decision to make. The BBC people we worked with never said anything to us, but we always had the impression that Clive Sinclair was lurking somewhere in the background trying to derail everything.

"Hermann had a most unusual approach when it came to setting salaries. He would say, 'Hey, let's go get a cup of coffee.' We went to the kitchen where he would say something like, 'We need to decide what we are going to pay you. What's a good number? Why don't you pick a number?' I suggested six and a half thousand pounds. Without batting an eyelid he said, 'Great, let's do that then.' I had spoken to some other potential employers before starting at Acorn. They weren't offering anything near that figure, so it was a really, good deal.

"Others who started around the same time as I did tell me they had the same experience: You'd go to the kitchen with Hermann. He'd ask how much you wanted. You'd say, 'this much', then he'd say 'yes'. That's all there was to it. In retrospect, it was a clever way to get people on board, happy and ready to start. Hermann and Chris always showed respect to everyone in the team. It was one reason why we were all so committed to them and to our work.

"Most hardware and software engineers working at Acorn were building the BBC Micro. They wanted to add an Econet interface but first they needed serial communications. When I started Joey (Joe Dunn) spent some time getting telephone communications working reliably. If you've been in a room with an acoustic coupler operating, you'll remember the infernal whistling that goes right through you. Imagine having to work in a really cramped office space, trying to concentrate, and there was Joey trying to get a cheap acoustic coupler to work with poor signal quality from a telephone system that was almost as old as our city's sewerage system – which was probably built by the Romans. Somehow, he managed to find a way to make it work.

"There were only three of us in the beginning, trying to improve the reliability of the Econet: Joey, me, and a guy called Steve – a young ginger-haired engineer who came from the north of England. The Econet used the HDLC/ADLC communication standard, invented by IBM and implemented as the Motorola 68B54 integrated circuit. It was capable of more than it was being asked to do. The signals were carried over twisted pair telephone cable. Being aimed at schools, which are always strapped for cash, everything had to be as cheap as we could make it.

"Paul Bond developed the low-level protocols and primitives. They worked fine for small networks, but work was needed to support larger sites. Carl Dellar worked with Paul on the first generation of Econet file server. Joey had written the network client for the Atom, which had been launched in 1980. They were all talented engineers, but the limitations of the Atom and early disk drives restricted what they could achieve. This is usually the way technology develops: Start with whatever is available at the time, then make improvements as new solutions become available.

"Except in small networks, you needed an angel on your shoulder to keep early Econet systems going. It only worked when the stars were aligned. Most of the work that Brian Robertson did in the first couple of years was to improve reliability of the hardware and low-level primitives. When idle, the cables weren't set low or high, they were allowed to just float. We could get away with that when the network wasn't busy, as there would be very few collisions (two or more computers trying to transmit at the same time). To be honest, Paul's primitives were, well, quite primitive. It was as good as it needed to be with a few machines, but it could be slow and unreliable. This became a problem when larger networks were installed. It needed to work with many more machines, over much longer distances; so that's the problem I started working on, with Joey's assistance.

"I spent my early years at Acorn improving the Econet, working on the client software for the BBC machine, and improving the file server. It took quite a few years because the file server went through several changes, ending with Version 6. It seems obvious now, but someone decided that random access read/write operations on files was a good idea, instead of: Start reading from the beginning of the disk and stop reading when you run out of things to read. We had to add a lot of new functions that would be performed over the network. The Econet update was hailed as a major release. Today, no file system worth its salt would get close to production without support for random access data files.

"The evolution of disk storage moved ahead in leaps and bounds. I know it sounds laughable now, but early 8" floppy disks stored about 100–200 kilobytes. They were expensive, quite large and difficult to store safely. The 5¼" disks came a bit later and could hold roughly the same amount of data. Then a couple of years after that, 3½" disks arrived. They were physically smaller again, but more robust. The quality of their media was generally better too. The data storage capacity of floppies doubled several times as the technology advanced from single- to double-sided disks, and from single to double, to quad density formatting.

"Finally, hard disks became affordable. The early drives on minicomputers had a fixed disk and a large removable disk pack. A common size had a 30MB removable platter on top of a 30MB fixed disk. They were known as the *Winchester 30/30* – inspired by the famous Winchester .30-30 ammunition of the *Old West*. Standalone removable hard disk cartridges were also released. With so many changes, someone had to be there at Acorn, updating driver software to allow users to hook up new storage technology. That was me.

"The first file server used the *cat command to read the directory from the first sectors of the disk. The single directory could hold only 16 files. It was replaced by hierarchical directories (like a family tree).

"During the early stages of BBC Micro development, we couldn't get enough disk drives, so we had to guard what we had. Our team only had three of them. If we went out, someone would come in and steal them. It gave a new meaning to the term: data security.

"Even for Cambridge graduates, who were continually breaking down technology barriers in the 80s, change that impacted our personal lives was greeted with some resistance. This was the era when the Iron Lady was tearing British industry apart, closing coal mines and car plants, and generally making a nuisance of herself. A lack of confidence was everywhere. Half a century later, we all seem to have adapted to living in an environment of constant change. I think what we were doing back then may have influenced that.

"We moved from Market Hill to Cherry Hinton before the silver building was completed. Most of us were still at university and rode bicycles to work. Few of us were fit enough to ride an extra six miles to Cherry Hinton and back, so it seemed daunting to us. To help get us all to work safely, Chris Turner arranged a meeting to explain how best to get there – complete with whiteboard presentation and printed handouts. Most of us were still wet behind the ears. He looked out for us, and we soon settled into the new digs.

"The Econet reached maturity about the same time as the first ARM chips arrived in April 1985. The ARX operating system was dragging its heels, so we started working on that with Carl Dellar's team in Palo Alto. Hugo and I worked on the file system, and we spent some time in California where we worked on ARX, though it was never finished.

"We were needed to develop software for things that would sell – like SCSI drivers for Archimedes machines. As usual, Acorn wouldn't buy it from someone else when we could make it ourselves! SCSI is supposed to be a standard, but then they released a second one. There were about seven major SCSI interfaces at the time, and we tried to be compatible with them all; so, we had to provide an interface that could support whichever 'standard' the attached device followed.

"It felt like the snowball was rolling down the hill, headed straight for us. With its incredibly complex specifications, being able to anticipate how the 'as-built' ARM CPU would perform was a colossal challenge. Perhaps our most important contribution to the development of the ARM chip was writing verification and validation code to test the first processor when it arrived. One weekend, Sophie wrote a special BASIC interpreter so we could build a tool set to emulate and simulate the ARM chip and its compilers.

"Writing the emulator enabled the hardware guys to test their chips and we built test software to ensure that the components behaved as it should. We needed to prove that everything worked correctly; that it would do what we thought it would do; and when we finally received the new chips, we could reasonably expect them to work. It would have been a disaster loading unknown software onto unknown hardware. What would we do if none of the lights came on?

"Many stories contribute to Hugo's reputation as a living legend: He was not above using a little profanity, which occasionally found its way into his software. For example, a barrel shifter is used to shift and rotate data words in the CPU within a single clock cycle. Hugo's validation code checked that a barrel shifter was working. It displayed 'OK' when it worked. We ran it against physical chips, and they always worked fine. Later, Arm sold the design to Digital, who made their own version of the ARM chipset. When they ran the validation tools, the barrel shifter failed. Their screens displayed the message: 'Barrel shifter fucked.' None of us knew about it. Before Digital made their own chips, it had never failed!

"That was around the time that Acorn ran out of cash. We did all kinds of things to keep the ship afloat. I moved over to Newmarket Road and that's where we started doing crazy stuff. They realized that the ARM chip had possibilities, with the Archimedes and Arthur being seen as the way to boost sales. They focused a lot of resources on them.

"In the end, I left Acorn because it was folding up. I was to be moved onto the NS32016 second processor. Someone asked: What can we do with 32-bit machines? If Acorn had a long-term strategy, it wasn't obvious. After a while I started thinking that there had to be somewhere else to restore the interest I had in the early days. Towards the end it was difficult to put on my shoes and go to work. It just wasn't fun anymore."

* **Note.** Developers were seldom known by their given names. Jeremy was Jes. Brian Cockburn was called Bruce. Brian Robertson was always BAFR. Joe Dunn was Joey. The nicknames just stuck.

00010100

Jim Mitchell

Jim Mitchell graduated with a degree in mathematics from the University of Waterloo. In the summer of 1965, as an undergrad, he led a small team to build a Fortran compiler on the University's IBM 7040 mainframe. It was used at 420 post-secondary institutions around the world. Optimised to perform fast compilations of student programs, WATFOR was about one thousand times faster than IBM's own compiler.

Jim completed his PhD at Carnegie Mellon, where he invented a compiler technique that is widely used today in many interactive languages like Python. He has since contributed to compilers, user interface design, interactive programming systems and language designs, document preparation systems, graphics hardware, distributed transactional file systems, distributed, object-oriented operating systems, the design of peta-scale supercomputers, microelectronics packaging, and silicon photonics for chip-to-chip communication.

In 1997, Jim was awarded the *J. W. Graham Medal in Computing and Innovation* in recognition of his contributions to the tidal wave of success that Java has experienced. This is how Jim describes his career…

"Between 1971 and 1984, I worked at Xerox PARC on a number of groundbreaking projects. In 1980/1981, I was Senior Visiting Fellow at the University Computer Laboratory where I came to know Andy Hopper, Carl Dellar, Hermann Hauser, and other Acorn people.

"In 1984, Andy, Carl, and Hermann convinced me there was an opportunity to build something great on top of the new RISC processor they were developing in Cambridge. Even though I had just accepted a position with Digital Equipment Corporation, their confidence and enthusiasm were contagious; so, we set up Acorn Research Centre (ARC) along similar lines to Xerox PARC – though with a much smaller budget.

"Brian Long, a fellow Canadian, was appointed Acorn's CEO in mid-1985. He didn't come from a technology background, but he was one of the best bosses I've had. He asked me if I would like to head up all Acorn Research, both Palo Alto and Cambridge. I was in my mid-40s. You might say there was some hubris, that led me to accept the role.

"I became President of the Acorn Research Centre and overall head of Acorn Research. From 1984 to 1988 I helped guide the Project A team in Cambridge that created the ARM reduced instruction set computer (RISC) chip, while also leading the ARX operating system team in California. If I had stopped to think about the air miles and being away from home for half of every month, I might have given it more thought.

"I saw my role as the R&D champion – to protect them at the Directors' level where budgets and other details were wrestled with, so they could get on and build the ARM processor without interference. It was a great privilege to work with the best and brightest at Acorn. I still think of Sophie and Steve as close friends – and of course, others too.

"One poor decision that paved the way with good intentions was setting up ARC in Palo Alto. To this day, there are people who question the wisdom of Hermann's, Christopher's and Andy's thinking. California was chosen to attract some of the best and brightest operating system experts who were leaving Xerox PARC in Silicon Valley. Since the main project was to create a new state-of-the-art operating system that would support the Acorn RISC processor it seemed like the right way to build the team. I now understand that the decision to create two organisations with competing cultures was an error that contributed to its failure; and that's too bad because ARX and other developments we worked on at Palo Alto could have been brilliant.

"The greatest challenge was managing the pressure from afar. There were people in the UK who were against everything we were trying to do. It could be seen in the petty games they played. It took a long time to understand the growing pains Acorn was going through and to work out how to manage them.

"As the funding crisis deepened, getting the resources we needed became a daily grind. For example, validating the ARM chip was not without its challenges. Acorn had nothing to run our simulations. It certainly couldn't afford to buy a VAX 11/750. We found a machine in Vienna where we could buy time at a price we could afford. So, we went to Hermann's homeland to complete the simulations. It seems kind of poetic, don't you think?

"Acorn's financial problems and near collapse followed soon after the formation of ARC. Some former Acorn staff describe development after Olivetti's takeover as: 'A hodgepodge of machines and system software'. They blame this on a lack of clarity about the business it was in. Others blame recruitment of middle managers from different areas of the computer industry for all the internal conflicts that festered there.

"The ill feeling and bad attitudes weren't present at ARC, where we were a small, but close-knit and supportive team. We took two years to build on what Xerox PARC had taken 10 years to develop. We were building on their and our experience. Since we already knew how to do a lot of what we needed, we almost achieved the impossible.

"Steve Jobs later demonstrated where we missed out. Steve's genius was packing it into a small computer with a Canon laser printer and a document preparation system. He figured out how to do that, starting with a system that could realistically be developed. He was superb at finding a price point that people were prepared to pay. The iPhone is another example of how he changed things.

"ARX implemented ideas that were at the leading edge of operating system architecture at that time. It was a multi-threaded operating system, and it supported multiple windows that could be shared over a network. Just imagine: People working over a network on different parts of the same document at the same time or looking at the same program. Back then, we were solving questions like: 'Whose mouse and keyboard has control when the window is shared?' Our solutions worked, though there are now better ways; but we were supporting features in the 1980s that wouldn't become mainstream until 20 years later. Who knows? Perhaps it was a bridge too far.

"Back then, computers were text based with a command line; yet with ARX, WYSIWYG was part of the design. Also, the file system supported write-once read-many (WORM) drives, which were higher capacity and cheaper than hard disk storage of the day.

"For a long time, BSD Unix has been compared to ARX; but it came along later with similar capabilities. That only happened after Steve Jobs was fired from Apple and he left to set up NeXT. It was some years later when he brought the NeXT team back to Apple that things started to change; and they had a modern operating system on which to build computers."

What I didn't learn at Graduate School

"I've learned that whatever you're building, computers or an organisation, if you want to succeed, you're best to start out small and simple. Build and prove your prototype. Acorn did that at first but failed to manage the transition. It grew too fast and couldn't cope with the change. Sometimes they got it right. Project A started with a small team and seldom had more than 10 people. National Semiconductor threw so much money and resources to develop the 32016. Big isn't always beautiful.

"Steve and Sophie have two of the brightest minds I've known. I admire the way Steve's training as a mathematician helps him focus his thinking. To him, every problem has an elegant solution. The ARM chip is so small, which is why it has achieved what it has. Fewer transistors mean lower energy consumption. That is perfect for the role that ARM chips now dominate."

What would I do differently?

"I regret not building trust between Palo Alto and Cambridge. At the operational level, we got on very well, but there were senior managers who plotted our downfall. And I regret not having the courage of my convictions before ARC was shut down. I should have convinced Olivetti to sell ARX off, rather than let it die. It was at least eight years ahead of the curve. Windows sat on top of MS-DOS back then. Microsoft hired a team of developers from Digital Equipment Corporation to build Windows NT, which was never as good as ARX – but it shipped.

"Olivetti could have sold the whole team as a package for a fraction of what Microsoft had to pay to DEC in an out-of-court settlement; or they could have set up another joint venture, as they did with the ARM chip. I blame it on a failure of marketing imagination.

"In 1988, I resigned from ARC and joined Sun Microsystems Laboratories in charge of research for Spring – a distributed, object-oriented operating system. Later roles included: Vice President of Technology & Architecture in the JavaSoft Division, Chief Technology Officer, Java Consumer & Embedded products, and Vice President Sun Microsystems Laboratories. I later became Principal Investigator on the High Productivity Computing Systems (HPCS) programme sponsored by the Defense Advanced Research Projects Agency (DARPA) and Sun. When Oracle acquired Sun Microsystems in 2010, I was appointed Vice President of Photonics, Interconnects, and Packaging at Oracle Labs. I retired from there in 2014. For some years, I've also held Board roles at the Curci Foundation, which funds research in the life sciences.

"Moving to Sun was an enjoyable experience, but I miss the times I spent staying at Andy Hopper's 16th Century mansion in Cambridge; stopping to buy cereal and other essentials because Andy didn't eat breakfast. I recall helping him cut out a section of the hedge at the end of his grass runway to make taking off and landing his aircraft safer."

00010101

Joe Dunn

Joe always appeared quietly confident, someone open to suggestions and feedback. Dealing with engineers on the other side of the planet, he never once complained when they reported Econet bugs, and always welcomed feedback. Joe treated everyone as part of the team. He built confidence to push ahead, even though it was never clear when the next bug might be uncovered to reduce customer confidence.

Joe Dunn's contributions to Acorn, both directly and indirectly, can't be overstated. His ability to engage with people draws on the way he trusts and is trusted by others. His achievements demonstrate that in the world of high-tech research and development, trust is a vital commodity. While soaking up the sun in Fiji, Joe took time out to share these stories…

"I was born in London. My parents moved to Canada when I was six months old, then moved back to the north of England when I was seven or eight. I grew up around Lancaster and went to Morcombe Grammar School. For my final two years, I went to boarding school at Seven Oaks in the South of England.

"I took a gap year, doing an internship for I. P. Sharp – a Canadian time-sharing, consulting and services firm. In 1977, I managed to get into Downing College at Cambridge, where I spent a year studying engineering. I was terrible at it, but towards the end of first year Cambridge started a two-year course in computer science. That's how I completed my Tripos.

"Looking back on my time there, I don't feel it was a great computer science course. It seemed to be part philosophy, part mathematics, and part engineering. Whatever it was, it didn't prepare people to become effective computer scientists. They taught programming languages but didn't teach programming techniques. If they did, I wasn't listening – things like how to do a bubble sort and how to build a database. I learnt a lot more practical computer science in a few months at Acorn than I did in my two years of face-to-face lectures at Cambridge.

"Nevertheless, I graduated in 1979 with a degree in computer science, then returned to I. P. Sharp for a while. My girlfriend (now my wife) Karen and I disliked being separated. One morning at a cash machine during a visit back to Cambridge, I ran into Mike Muller, a fellow computer science graduate. He said he was doing some work for a guy called Hermann Hauser who was looking for someone to build a 68000-based Cambridge Ring workstation. It needed to be done quickly. That sounded interesting so I met with Hermann. He talked about what he wanted to build. I soon joined Acorn, but Hermann changed the brief. 'We're going to build a file server.' That was the last time I heard about 68000 workstations.

"Hermann introduced me to Paul Bond and Carl Dellar. Paul knew everything worth knowing about the Atom and Eurocard operating systems. Carl was building the file server and was the Econet lead, so I worked for him with guidance from Paul.

"As I've said, my Cambridge computer science degree hadn't prepared me for what was to come. Unlike many of the early Acorn engineers, programming wasn't my hobby – an obsession that occupied my every waking hour. But here I was, writing client code for the Econet interface. They just gave me a 6502 handbook and said 'OK, off you go.'

"Having done no machine level programming, the learning curve was steep. In hindsight, it was tough to find my way into it, though it didn't seem so bad at the time. Not only did I know nothing about networking, I also wasn't a trained programmer. Sophie Wilson and Laurence Hardwick seemed a little dismissive of my programming ability – justifiably so. It just made me push myself to get better. It took some time for my coding to reach the level of the people around me. Fortunately, Carl Dellar was a brilliant leader and mentor.

"I started by writing Econet client code for Eurocard Systems, then compressing it to fit in the available space of the Atom. It had to fit in only two kilobytes of ROM. To achieve that, it needed a lot of compromises, but I managed to make complex functions including the collision detection and collision arbitration code fit into that 2k. With Carl's help, I developed a few tricks to make it fit. For example, I moved command line processing into the file server. Compressing it took more time and effort than writing the first version. It reminds me of that adage: *Had I more time, I'd have written a shorter letter.*

"The Atom and the System Econet client code were ready to ship by the middle of 1981. There were a lot of other things going on, but we kept improving it. Working conditions at Market Hill were difficult because space was unbelievably tight. I shared a corridor with Carl for a while. There seemed to be a lot of other activities going on around us; then one day we heard that Acorn would be doing the BBC Micro. It's not that it was a secret, it just happened so fast. Meanwhile Carl and I were busy plugging away at the file server.

"Not long after the BBC announced their decision, Hermann said, 'Hey, we're going to need a new operating system for the Proton.' Next day, I started working on the Beeb's operating system under direction from Paul Bond. I started on the cassette filing system. Gradually more people came to work on the new machine.

"Paul finished MOS version 0.1, then said, 'OK, we'll let it run.' It couldn't talk to the Econet or disk filing systems, but it allowed us to ship the basic machine. The Beeb's printed circuit board was designed to take the hardware, but we didn't have a MOS that could talk to extra filing systems until early 1982. The Disk Filing System (DFS) came first, then the Network Filing System (NFS) followed some months later.

"I remember spending countless hours in discussions about the project. I was up late at night fixing code to get it ready. I do recall when the BBC guys came around. We were trying to finish the final code, and I was busy taking bugs out of the MOS; some were just stupid mistakes that reflected the pressure we were working under at the time.

"When I wrote functions in the Atom, we had no room, so we took commands and moved them to the file server, which sent back the code to execute. For Atoms and Beebs to run on the same network, the code had to be functionally the same. The BBC Micro had its own command line primitives in the MOS (star commands), so it performed its Econet client functions like the Atom but called them through the MOS.

"All BBC filing systems – cassette, disk and network – are extensions of the MOS. Applications pass data and commands through the MOS. For example, if you're in BBC Basic and you send the *cat command to get a directory listing, the interpreter passes that command to the MOS, which in turn passes it to the active file system. This could involve reading a cassette tape, a floppy disk, or passing a request via the Econet to a file server. Paul Bond wrote the low-level code in the MOS.

"Carl Dellar was moving to the USA, so they decided to put me in charge of the Business Systems group – I guess I was last man standing. We were still in Market Square. It was tough because I was young and had no experience with project management. Regardless of what I thought of my programming skills, I knew I had to step up and be the man.

"It was also the case that I was a very bolshie kid. Early on, I tried to understand why it was so difficult to get software done. I often formed strong opinions and wasn't afraid to share them. Maybe Hermann picked up on some of that.

"Thinking about it now, my job was quite lonely for someone so young and inexperienced. Acorn was a small company filled with a lot of inexperienced people. None of us had done anything like that before. We didn't have the Internet. There was no Acorn forum where we could see what bugs people were finding. Anyone who showed an interest and who seemed to be intelligent was a good person to talk to. Besides Rob Napier and Brian Cockburn in Australia, there was Paul Bond and Kim Spence-Jones. Jon Thackray took over from Paul later, and there was Laurence Hardwick, Brian Robertson and Andrew Gordon.

"I don't remember working with Sophie much, just occasionally on bugs. Paul would sometimes come in and do low-level code, then I'd add the higher-level parts on top of his. Paul also worked on parts of the system like the voice processor. I'd sometimes add code on top of his, but it was trivial. I do remember discussing additions to MOS 0.1 with Paul and getting work done to enable disk and Econet filing systems to work.

"My main contribution to the BBC Micro was Econet. I well recall writing cassette functions because I often stayed up till 2:00AM getting bugs out. I did the filing system on top of Laurence's work. If you want to write a file you have to tell the tape to start, tell the tape to write, and then tell it to stop. He did the underlying stuff.

"Some months after the disk filing system was operational, the Econet started to work too. I remember the order of getting things done because we had to use the low-level parts of the DFS to write the file server.

"It was early 1983 when the Lisa came out. It just blew my mind. I went to the US for a job interview with Intel. I wanted to move to the States, and I went into a Palo Alto computer store that had a Lisa on display. It was absolutely mind-boggling. It was like nothing I'd ever seen. When I went back to England I badgered the guy in charge of the software group. I'd say we had to get a Lisa. Finally, he agreed, so I called around until I found an Apple dealer somewhere in the South of England that had one. We all jumped into our company cars and zoomed down there. I put it on a credit card, took it home, then Karen and I played with it all weekend. I brought it into the office on Monday morning. Sophie took one look at it, then disappeared from the office – with Lisa. When she re-appeared a week later, she announced: 'This is cool. This is how an organized system should be structured.'

"Suddenly we were in a different generation. There were some nutty ideas about how we should build a machine like that. It was very exciting with all that bubbling around in 1983, along with the idea that we should build a machine based on a 32-bit processor.

"Mid-year, I visited Barson Computers had me come to Australia to promote Acorn and the Econet. I really enjoyed that trip. It was nice to see our work was appreciated."

INSIDE THE ECONET - AN ADVANCED NETWORK USERS WORKSHOP

Tuesday 9th August

As a complete Econet novice, it was with some trepidation that I took up the challenge of a workshop for "Advanced Network Users". I found myself crammed into Barson's conference room with about 20 other enthusiasts, some from as far afield as Queensland and South Australia. The workshop was ably conducted by Joe Dunn, project leader of the Econet development group.

Joe's visit to Australia was prompted by the frequency and length of the telephone calls between Barson Computers and Acorn. They decided that it might be better for him to see the development work that Bob Napier and Brian Cockburn have put into Econet, and where the majority of BBC Micros running Econet are installed!

Joe is youthful and blonde haired. He looks to be the epitome of a Cambridge graduate, but he may well have been educated at Battersea Polytechnic, for all I know. He spoke authoritatively, without any sign of jet lag, despite arriving from England barely two days earlier. For nigh-on six hours, he explained the intricacies of packets, receive and transmit, four-way handshake and other mysteries.

I was interested to hear about the Econet's design philosophy. To keep costs down, off-the-shelf components were chosen, such as the RS422 line drivers and the 6854 chip that defines the data packets. The result is that the cost of an Econet interface is about $200, with the cost of an Omninet interface is nearer to $1000.

The workshop showed how to write software that enables networked machines to communicate with each other, using the network primitives – especially receive and transmit. All I need now is a network to practice on. My personal single, cassette-based, Beeb isn't quite up to it!

John de Figueiredo
AEUG Vol 1 Number 4
Newsletter of the Acorn Education User Group
September 1983

ARM development

“Late in 1983, Steve, Hermann and Sophie were trying to figure out what processor to use for the next generation of machines. Hermann said to me, ‘I’ve started a group.’ I was astonished when he invited me to join his brains trust. I was hardly in the same league as Steve and Sophie. It was nuts because I didn’t have the intellectual heft to be in that room. But I was, and party to some of the conversations around the fact that we didn’t want to use the 68000, didn’t want to use the 32016, and certainly didn’t want to use an Intel 80286. OK, so what do we do? That’s where the germ of an idea for the ARM came from.

“Hermann came back from one of his regular trips around the world. ‘I’ve hired a guy from Xerox PARC to start up our office in Palo Alto. It’s fantastic in California. You can get things done in a week. You can just rent an office and go to people and ask if they want to change jobs. They agree, and then you set them up.’

“It was true. Hermann had persuaded Jim Mitchell to set up Acorn Research in Palo Alto. Jim had been one of the brains behind the successes at Xerox PARC in its glory days of the 70s and early 80s. He designed file servers and developed one of the first high-level systems programming languages. Acorn would build a machine that ‘knocked off’ all the best Xerox ideas. We weren’t going to use Unix. We’d build a multi-threaded operating system with a Window front-end; we would build a multimedia editor on top of the operating system; and it would have a new and innovative log-based file system. We were going to do all of that ourselves. This was all very exciting because that was exactly the technology I wanted to do. I just thought it was fantastic.

“More experts were recruited. Jim hired some people with English PhDs who had already moved to California. Carl Dellar was there while I was in Palo Alto, along with Jonathan Gibbons and Randy Strauss – operating system superstars. We started working with this new group. There was a genuine belief that it was going to be amazing and magical and very Xerox PARC, but it would be ours.

“There were cultural and technical differences between Palo Alto and Cambridge. Our people in England are great hackers. They were fast with low overheads in machine code. But they would have to connect their work to the software written in Modula-2+ by people with PhDs in advanced operating system design, who wanted to build a large yet-to-be defined generalized system. Where’s the documentation describing what this is going to do?

"I'd accepted that I wasn't strong technically, but I could organise people and wasn't afraid to share my opinions. I spent 1984 trying to make sense out of what was happening. I handled project management for Jim. Eventually he said, 'Why don't you come out and we'll get you a green card in a year or two.' That's why I ended up in Palo Alto.

"I loved the PARC concept and thought it incredibly visionary and exciting. I was in my twenties. I thought that this is how machines should be made. It was different from Cambridge, which wasn't doing any of that stuff. It was incredible technology that nobody else was building; and there was the idea that we could build things that were beautiful – like the Lisa. There was also the idea that if you had computers that were easily accessible like that, you could advance the human condition. We would all be better for it; so that thinking was very exciting. It was wonderful. I watched brilliant people like Sophie, Steve and Hermann wrestle with these ideas. They had different ways of looking at the same problems from different technical perspectives.

"Jim Mitchell was absolutely fascinating to watch in action. He came from Xerox PARC, which took the view: 'We're going to build it all ourselves and we're going to build it on machines that are powerful because we know that computers become more powerful each year. Let's build for the future. We'll see the machines being done right and we'll write everything in high level languages. That seemed to be Jim's point of view. Everything should be elegant.

"Meanwhile, Sophie wanted to build everything from scratch. You make it run by taking ideas such as text and windows systems, and you make your own versions. OK, beautiful; very fast, stripped-down versions of those things; and we had very different ways of doing things. That was the difference between Silicon Valley and Cambridge.

"Karen and I had already decided to move to the USA before this happened. I learnt that I could live in San Francisco and work in Palo Alto. We flew to California in July 1985.

"In the middle of this move, Olivetti took over Acorn. Fantastic! With only an H1 visa, I learnt in my first week after getting there that my dream job was on the line. It looked like my chances of getting a green card were slipping away.

"I spent 1985/1986 trying to help Jim and the team develop the operating system. I corralled people on both sides of the Atlantic. I went to board meetings every three months and presented to the board – always hoping they would buy it.

"In my view Acorn sat in limbo for a couple of years trying to decide what to do with ARX. Hermann may see it differently. Acorn was owned by Olivetti, and they were trying to figure out what to do. There was one hilarious trip that I took with Jim to Ivrea, Italy, where we had an audience with señor Pioli, Olivetti's CFO. At one point he said, 'Jim, every year I must go to Microsoft; and I have to say Bill, what do you have for me? I must accept what Bill gives me. So, what could you give me that will separate me from Bill?'

"That was their agenda: They thought they had some California magic that could be a breakthrough for Olivetti. Señor Pioli had two fingers missing in the middle of his hand. At one point he held his hand up and motioned to stop. It looked threatening. I didn't have a green card at that time, so whatever decision he made, it stood to affect me personally. If we failed to impress him in this interview, he would shut the operation down and I'd have to return to England, which I really didn't want to do.

"Jim just started doing a sort of fantastic tech shuffle. He threw out a stream of buzzwords that completely lost Señor Pioli and me: 'ARX has an object-oriented system that has a great UX…' Using words like UI/UX in those days blew people away. He talked for about 10 minutes. I have no idea what he said, but then, neither did Señor Pioli, who couldn't bring himself to admit it.

"We came out of that meeting in good shape. I have no idea what we promised to give him. I think we may have promised to finish this amazing operating system that could cut out Microsoft Windows. I guess at some point there was the idea that it would run on all the best computers; or maybe the ARM chip would become a winner, which Olivetti owned at that time. Maybe they would start building great computers with ARM chips and this amazing operating system. I haven't been in a conversation like that before or since.

"We could have gone with Unix, but it was clunky and slow and wasn't multithreaded. Back then, it was just big and clunky. There was Berkley Unix, which Apple went with, and which did a great job. Apple always seemed to have 'the touch'.

"I don't remember the details now but so much was happening with operating systems. Window systems were absolutely cutting edge. Nobody knew what they should look like, but everybody had very strong opinions about what was good, and what wasn't. They differed on how to build a windows system – even what it meant to have one. Is it going to be fast, or clunky and slow? If you build it on top of Sophie's custom-built operating system, it's going to be fast.

"You want everything built from the ground up. You don't want a big slab of Unix underlying your beautiful infrastructure. That discussion was everywhere, much like AI is now. People say: 'You can do this; you can do that.' But we don't really know. The bean counters want instant results but let's face it: The big jumps in technology don't happen overnight. Apple came out with the Macintosh in 1984, but it took another 20 years for it to become an overnight success. The problem with Acorn's work in Palo Alto was that Olivetti wanted to see results in 20 weeks.

"When the ARM shipped, it was a generational breakthrough; but the ARX operating system was massive and inappropriate. It would never succeed because it was beyond our ability to deliver – though no one wanted to admit it. It's fascinating at many levels: How difficult it was to recreate Xerox PARC and make it work. We failed to do it, but the Macintosh team succeeded.

"During that period, we had a guy running Acorn who came from the tractor business. He was a nice guy and clearly a professional manager, but he was being presented with two or three options like: Should we use Unix? Should we use this crazy thing they're building in Palo Alto? Should we use an operating system that Sophie and the guys in Cambridge want to build? He wasn't qualified to make that decision. He didn't know how these things went; so, there were board meetings discussing this question.

"By that time, it was clear that the operating system wasn't going to work. It was just too big, and besides it was getting rejected by the English group. I received my green card in 1987/1988 and decided it was time to move on. I could now work anywhere in the States."

After Acorn

"After leaving Acorn, I went to work for Frame Technology. They were a genuine Silicon Valley startup employing 40 people when I started. They developed *FrameMaker*, which is a great application for writing technical documents; now owned by Adobe.

"It's interesting to compare the difference in approach between Acorn and Frame. The American outlook is quite different: The goal was to get the code out. You sell it, make the stock go up. My job went from wrangling PhDs to writing documents that stated what we planned to do. The other was managing deadlines: What must we cut so we can ship on time? Why is it going slow? It's the sort of thing that people with a grip on reality do to get the job done.

"I took six months off in 1990 after Frame Technology; then I joined Macromedia, best known for *Dreamweaver* and *Flash*. That was a fantastic fit. I loved the culture of it. I loved the stuff they were building; and I was there through the 90s till I tried to retire in 2000.

"That lasted for about nine years. But I couldn't stay retired, so I started executive leadership coaching of CEOs about 10 years ago. I'll keep doing this as long as I am challenged by it."

00010110

John Cox

Just before his eighteenth birthday, while waiting to start first term at the University of Cambridge, John Cox joined Acorn Computers. During his eight short months at Fulbourn Road, John ported the Eurocard System disk controller software to become the BBC Micro's Disk Filing System (DFS). This is John's story…

"My father was a civil engineer. Born in Devon and raised in Surrey, I come from a middle-class background. There was a short stint in Jersey at the age of two, though of course, I'm too young to remember any of that. At an age when most boys dream about football and skateboards, my parents encouraged me to be studious and to try my hand at whatever caught my fancy. My main interest was electronics: resistors, transistors, integrated circuits and all that sort of thing. By the age of 10, I was experimenting with electronic circuits, though some of my early attempts to fix things resulted in total and permanent destruction of the item. But that didn't deter me (or my parents). By the age of 15, I had built a microcomputer. It's fair to say that I was mostly self-taught, learning by reading and experimenting. Those habits have stayed with me throughout my career. I feel I'd been preparing for my job at Acorn since I was a child.

"In the mid-1970s I received a scholarship to attend Oundle School in Northamptonshire. My parents and I had visited a few good schools, but Oundle offered the best opportunities to get more involved with computers. If you've ever been to Oundle, you'd know that there wasn't much to distract me; and besides, the school didn't go co-ed until 1990. I spent a lot of time honing my programming and hardware skills with the school's Data General Nova minicomputer and the Zilog Z80 microcomputers, helping to design and build a Motorola 6800 board, and building a 6800 computer from a kit.

"John Coll taught electronics, though I didn't have a lot to do with him back then. He had been working with the BBC and Acorn, writing the user guide for the BBC Microcomputer. He knew I had worked on a Z80 project at Torch Computers in Cambridge over the summer that year, while preparing for my A Levels. In September 1981, I mentioned to him that I was looking for a job until I started my computer science degree studies the following year. With John's recommendation, Sophie found me a job porting the Eurocard System DFS to the soon-to-be-released BBC Micro. I worked there for nine months, until I started university.

"John left Oundle School and later joined Acorn Computers as Manager – Education. He had a house in Cambridge and invited me to stay with him for the first couple of weeks until I found a place of my own. I then moved into some rooms at the old Castle Hill Police Station that Andrew Gordon arranged for us. I lived there until I started at Pembroke College.

"It was the Acorn way to just throw you in the deep end and expect you to start swimming – or to be more correct: start reading. I knew Z80 and 6800 Assembler, but I hadn't worked with the 6502. It was a simpler chip, so it didn't take long to learn. I was given a desk in the office next to Sophie's, a copy of the *6502 Assembly Language Programming Guide*, a populated BBC Micro circuit board with on-board disk hardware, a disk drive, monitor, power supply (Allen Boothroyd was still developing the case.) and a source listing of the System DFS. I set to work.

"The Intel 8271 floppy disk controller was not a complex device. I understood the hardware and its very simple commands: *configure*, *check status*, *format*, *seek*, *read*, *write* and *delete sector*. I didn't need to bother Sophie much, but she was always happy to help me when I needed it. Paul Bond had completed his first release of the Machine Operating System (MOS 0.1). The DFS doesn't have many interactions with the MOS, but a few changes were needed to get my code working. Paul made those changes for me in MOS 0.2.

"Things must have moved quickly after that, because I remember taking my Beeb home for Christmas with the disk drive working. Well before Easter 1982, we had MOS 0.3 or 0.4 and DFS 0.2. We had disks working well enough to release for external testing. This is the version that was sent out to Barson Computers as a prototype for the Western Australian Education Department tender evaluation.

"By the summer, DFS was on the market. My work continued as I refined the software until I left in the autumn of 1982. Before leaving, I spent my last few days at Acorn building my own BBC Micro from parts that were lying around – and the only black BBC Micro case known to exist. I moved to Pembroke in October to start my first year – mathematics, followed by two years of computer science. I returned to Cherry Hinton a few times to fix some DFS bugs after I started at university.

"Reflecting how fast disk technology had advanced in only a few years, the first DFS had a very crude flat directory structure. This wasn't a big problem on a 100-kilobyte drive, but by the time quad density, double-sided floppies came along, and then hard drives, something better was needed. Hugo Tyson did a great job writing the Advanced Disk Filing System (ADFS). He also replaced the ageing Intel 8271 with the Western Digital 1770, which supported the higher density drives.

"While at University, I worked part time with Andrew Gordon and Kim Spence-Jones. Having my own BBC Micro to continue learning on proved to be useful. Besides my work for SJ Research, which was heavily focused on the BBC Micro and Econet, I wrote the BIOS for Acorn's Z80 second processor.

"In both my life and work, like many others who started with Acorn Computers in the 1980s, I haven't strayed far from the nest – geographically and metaphorically. I joined SJ Research after graduating from the University of Cambridge. My work with Kim and Andrew was always enjoyable and professionally rewarding. There never seemed to be a reason to go elsewhere.

"I wrote the Hard Disk Filing System (better known as SJ's HDFS). As a project for Acorn, I worked on an Ethernet Card for MS-DOS. When SJ Research started working on X.400 messaging, I was there working on that development. I also spent some time on network packet switching.

"For many years now, I have been working with video decoder technologies. I believe I can recite international standards, chapter and verse. My current work is consulting for the Raspberry Pi on video decoding. I expect that this will remain my focus for the rest of my career, though I have no plans to stop any time soon."

00010111

Jon Thackray

Jon Thackray joined Acorn Computers at the beginning of 1982, having completed his studies at the University of Cambridge. He was hired in late 1981 to replace Paul Bond. Jon would lead development of all future releases of the Machine Operating System (MOS), as well as assemblers, debuggers, emulators and a successful adventure game with Acornsoft. At Acorn, life was never dull. For example, he appeared as expert witness in an Australian copyright dispute. Here are Jon's recollections of his time at Acorn…

"I don't believe in predestination but looking back at my interests as a callow youth in Surrey at age 14, it does seem that my route to Acorn started there. My school friend, Mark-Eric Jones' dad had a business in computing, so MEJ had a computer in his bedroom, which was quite unusual in 1969. I don't remember much else about it. I think it was just inspiration for what followed in sixth form. MEJ learnt Algol 60 and I studied macro assembler on an Elliott 803 with its 8192 words of 40-bit ferrite core memory. It lived in a room about the size of a large living room. This monster machine was great fun to explore.

"MEJ and I both went on to Cambridge – he studied engineering and had a successful career with at least two international patents to his name. I chose mathematics, which I find comes in handy when it's time to balance my chequebook.

"During my first year at university, I asked MEJ if he had any interesting computing problems to solve as I'd been starved of programming. He told me that his father had started a company to produce a word processor, based on a new chip called the Intel 8080. They needed a ROM-based paper tape loader. It was early days for the 8080 so the only documentation we had was an instruction manual for its predecessor, the 8008, with an addendum describing the new instructions. I wrote the loader in hexadecimal, as there was no assembler or other development tools available. I also wrote an 8080 emulator to test my loader on the university's IBM mainframe – so started my lifelong interest in emulators.

"After completing my undergraduate studies, I went on to complete my PhD. I didn't spend all my time studying; I managed to waste quite a lot of it solving a game called Colossal Cave. It was there I met fellow PhD student David Seal. We wasted even more time writing a compiler and interpreter for building adventure games. We then co-wrote Acheton, later published by Acornsoft. David and I ported our adventure game generator to the BBC Micro, and later to the Archimedes. David became Acorn's in-house graphics expert, taking over from Graham Tebby, who had written the original code.

"After completing my PhD, I spent 18 months helping the mathematics department publish a book called *An Atlas of Finite Groups* – which is great bedtime reading (and less addictive than sleeping pills). As that work was coming to an end, I managed to get a friend to introduce me to Andy Hopper, who had been his lecturer in third year computer science. Andy had been in business with Hermann Hauser at Orbis, developing the Cambridge Ring."

My Start at Acorn

"By then a Director of Acorn Computers, Andy recommended me to Hermann, who mentioned me to some of his software team. One of them was Paul Bond, a consultant with TopExpress, who had written BBC MOS 0.1. I'd met Paul in the Computer Lab, when we were doing our PhDs. Acorn wanted to replace him with a full-time employee who would maintain the MOS over the long term. I was offered the job of managing development for versions 1.0 through to 1.2. Initially, I worked alongside Paul.

"The concepts behind the MOS have received a lot of praise. Those of us who worked closely with it, know that Paul developed the architecture according to good computer science, adapting ideas from the Acorn Eurocard Systems and the Atom. That takes nothing away from his work. There are far too many software engineers who ignore good design practices for expediency. We have all benefited because Paul is not like that. After becoming familiar with the MOS, I took over from him.

"MOS 1.0 wasn't developed in the way you might expect, starting with a specification and working toward a completed product. Due to the way the Beeb's hardware developed, we dealt with bug reports and change requests as they came in – always looking for places to reduce code that made room for changes. Over time, the reports and requests became fewer and less frequent. One day, Sophie Wilson and I agreed that it was time to let MOS 1.0 loose on the world, and we sent it off to make a mask

ROM. Unfortunately, a few obscure bugs raised their heads and before we knew it, MOS 1.1 came and went, then finally MOS 1.2. After that, we had a stable version that stood the test of time."

Tools, Tools, Tools

"The Universal Assembler/Disassembler/Editor (UADE) and the Universal Assembler (UASM) in use at the time were completely inadequate for the task, so building on my experience writing the 8080 assembler and the adventure game compiler, I rewrote the 6502 assembler, which became accepted as the tool of choice. I later developed the Acorn Macro Assembler (MASM) for the 6502 second processor; then super-sized it for the Turbo 6502 second processor.

"I collaborated on development of a 6502 toolkit, including macro assembler and debugger. Others added editors. Through gradual evolution and improvement, MASM became the ARM assembler. I seemed to be destined to keep writing newer and better versions of the same program. This included the first instruction level emulator for ARM.

"Steve Furber had written a gate-level emulator for the ARM that ran at about three cycles per second. It wasn't much good for running code on, but it did what he needed it to do. I wrote the first ARM emulator to run on a 6502 second processor. Mine ran at a couple of thousand instructions per second. Again, it achieved what I needed out of an emulator. Sophie wrote an ARM emulator on the 32016 second processor. She also implemented an editor called TWIN (meaning Two WINdows). It was remarkably useful."

Rock solid

"One Saturday, I was working in the silver building, affectionately known as the *Tin Palace.* As usual, my BBC machine had no lid, so the board was completely exposed. Distracted for a moment, I spilled a cup of coffee on it. The mains circuit breaker tripped on the other side of the building, shutting down all electrical equipment near me. Eventually, I managed to find and reset the tripped circuit breaker. After drying the computer out, I was surprised to find that it powered up and worked fine – what a robust machine!"

False Economy

"There's no escaping it: At some time in our careers, we are all forced to deal with the mundane. For me, it was the Purchasing Department's decision to buy cheap Wabash floppy disk media. Floppies were the lifeline by which software was shared – in and outside the office. Most floppy disks had limited life, and some were unreliable straight out of the box. Some brands were better than others: From a box of 10, it wasn't unusual for two or three Wabash disks to fail. Yes, they were cheap, but complete rubbish. We had a hard time persuading purchasing that buying them was false economy. Thankfully, the days of floppy disks are behind us. They have gone the way of the steam-powered traction engine and the 8-track tape player."

The Electron

"Christopher Curry believed there was a market for a cheaper less-capable BBC Micro, so the idea of the Electron was born, with me in charge of MOS development. To ensure an unofficial version of the MOS wasn't used in production by mistake, we had a few screen names in place of 'Acorn Electron'. My favourite was 'An electric elk called Simon' – a reference to a Monty Python sketch about an author called Nevil Shunt. This is how the Electron's nickname came to be the 'Elk'.

"Several high-cost discrete chips such as the UART and display controller were to be replaced by an Uncommitted Logic Array (ULA) from Ferranti. It would also use a slower clock, which presented its own side effects. Acorn already had some experience with ULAs, having designed the video, serial/cassette tape and Tube ULAs for the BBC Micro. Nevertheless, the Elk ULA was to be the largest ever attempted, needing around 2,000 gates.

"We had no simulation tools back then, no synthesis tools, no clock domain crossing (CDC) checkers, or anything else that might have made the task seem less like landing a man on the moon. All we had was a layout tool that was a large BBC Basic program written by Sophie. The lack of tools and and too many fabrication problems contributed to the delays.

"But added to this, we didn't appreciate at the time how little Ferranti understood about their own products. This proved to be catastrophic for the future of Acorn Computers as a viable business. After six reworks of the ELK ULA, the product missed the 1983 Christmas market, almost sending the business to the wall."

BBC Model B+

"While the Elk was causing Acorn's directors to age prematurely, another storm was brewing. Hitachi announced plans to stop making 4816AP-3 (16k x 1-bit high speed DRAM chips) as used in the BBC Micro and that the smallest size would be 64k by 1-bit. The original design couldn't be built using these new chips, so the idea of the Model B+ was born. It could now have more memory for graphics and not penalise programs using screen Mode 0, which needed 20 kilobytes of RAM.

"Meanwhile, we tried to sell the BBC Micro in the USA. This suffered the same fate as most UK companies do when they try to branch out there. People were recruited who seemed to spend their time flitting around the world, but nothing ever got done and never went anywhere. Acorn spent a lot of time and money dealing with the fact that the US has much tighter radio frequency emanation rules. Also, television in the UK used a PAL standard at 50Hz mains power while in the USA, they have the NTSC standard at 60Hz. Resolving the technical problems with the US/CAN BBC Micro and the Electron cost Acorn its future."

Broken Break instruction

"Around this time disk controllers were being added, along with other extensions, such as extra memory and Econet interfaces. We noticed a problem with the 6502 microprocessor: The assembly language BRK instruction had been used to flag errors, followed by a text message ending with a zero – which is the machine code for BRK. The Econet interface sent non-maskable interrupts (NMI). If you aren't familiar with NMIs, think what happens when your wife/husband/partner/significant other tells you to take out the rubbish. It doesn't matter what you're doing at the time, you are expected to drop everything and take out the rubbish. In CPU-speak, an NMI is asserted. Compare this to a maskable interrupt request (IRQ). That's one where you fit your noise-cancelling headphones and block all external requests until you are ready to listen.

"We found that the BRK instruction could be interrupted by an NMI but would restart one byte later. This led to a lot of rubbish (the error string) being executed as instructions, and then the terminating byte set to zero at the end of the string being treated as a BRK, thus producing an error message that displayed rubbish. Fortunately, we had changed to the CMOS version of the 6502, which treated unknown instructions as NOPs, rather than BRKs. The problem went away.

"I spoke to Rockwell, the manufacturer of the 6502 (along with Synertek) about this problem, and was told, "Oh yeah, we know about that." We wondered: 'So why didn't you %$^%$# tell us?'"

Master 128

"In the mid-1980s, we needed a larger, more capable BBC Micro, so the Master was born. It had multiple ROM sockets to meet the growing demand for more ROM-based applications software, while using the same memory layout and screen memory allocation as the Model B+.

"Another feature was an on-board real-time clock, which needed a small amount of non-volatile memory and a battery to keep it running when power was disconnected. This led to the infamous *exploding Master problem*. Batteries would sometimes overheat due to overcharging. In one case, a fire started in a hospital where a Master was being used as a data logger. This caused a serious PR storm for Acorn. It seems people disapprove of hospitals burning down."

ARM

"Hermann had visited California in 1983 and came back talking about RISC machines. We'd always been unhappy with the 32016's slow interrupt response (latency). Steve and Sophie saw this as an opportunity to design a chip that would be fast, with a low transistor count due to its reduced instruction set. We needed a microprocessor with an interrupt latency of a couple of microseconds. Once they realized they could do this, the Acorn RISC Machine (ARM) chip – Project A – was born.

"By 1984, Acorn was in financial difficulty. The Electron's manufacturing delays left many thousands of machines sitting in a warehouse with nobody wanting to buy them. Olivetti acquired about 79% of the company but for a while, things continued pretty much as they had done.

"The ARM1 came back in 1985. It was almost perfect. Not long after, the VIDC, MEMC and IOC support chips arrived. We had a complete chipset to build computers. Sophie and Steve could see what success looked like. Those first samples were widely considered a great success. It did have a minor fault but that didn't reduce demand for it – both inside and outside Acorn. The flaws were fixed, new features added, and notable speed improvements made – all part of the first production run (ARM2), by which time the other chips in the ARM family were also available. It was a success and used as the basis of a new series of machines – the Acorn Archimedes. ARM3 came along some time after that. It was the last one that Acorn made before ARM was spun out into a joint venture with Apple and VLSI Technology.

"Over the years, comparisons have been made between ARX (developed by Acorn Research in Palo Alto) and RISC OS. During development, there was talk about making a rival to the Apple Lisa and Macintosh. I even went with David Lamkin to buy a Lisa. We loaded it into the boot of our car and skulked back to Acorn. The sense of guilt for having a competitor's product was palpable. We liked to think that we didn't need to copy anyone, but this adventure all turned out to be a waste of effort as it was clear that it would be much too expensive to produce; so, we took an alternative cut-down approach starting with *Arthur*, which evolved over time to become RISC OS. I worked on various parts of this, notably the Task window."

Copyright Violations

"I was not formally trained in computer science or engineering, but during my time at Acorn I learnt a lot through hands-on experience – things such as 'race conditions' and 'critical regions'. One of the more amusing turned out to be in the disc controller. To avoid wearing a slot in floppy disks the read/write head lifted off the disk after a delay.

"There was a critical period of about five milliseconds before lift-off when it could accept commands but would be unable to perform them due to the head lifting. Hugo Tyson and I wrote a program to test for the existence of this bug, as it indicated that the code had been copied from our original DFS – and believe me: There were a lot of pirated copies. We spent a day wandering around a computer exhibition looking for tell-tale signs of this, noting which vendors were selling pirated code and which had developed their own bugs.

"Acorn's largest international distributor was Barson Computers, which sold a lot of BBC Micros to Australian and New Zealand schools. Government schools purchasing was organised on a state-by-state basis. In Western Australia the government tender rules changed after 1982, when the Beeb was first approved there. The Education Department could no longer specify which machine must be purchased; they could only specify its operating characteristics. When the tender came up for renewal in 1987, they effectively specified a BBC computer without mentioning it by name.

"Two companies tried to make a new machine called the Squirrel that was a copy of the BBC Microcomputer. Professional Technology had the technical people and Southern Technology drip-fed cash to fund the enterprise. It should come as no surprise that Acorn and Barson were up in arms about this. They filed lawsuits and injunctions to stop it.

"The Court needed an expert witness to examine the Squirrel and to decide whether it was a copy. Amazingly, Barson's lawyers had convinced the judge to allow someone from Acorn to come and say whether it was a copy. One day in October 1987, Sophie came in and said, 'We've got a problem. I want you to go to Australia for a month.'

"This was by far the longest flight of my life. It took 24 hours to go halfway around the world in a Boeing 747. I arrived in Perth late in the evening, dead on my feet. I was met by Barson's representative who took me straight to my hotel. I felt like a zombie.

"When he collected me the next morning, I didn't feel a lot better. The lawyers had obtained access for me to examine the defendant's code. We visited the offices of the company that had ripped off Acorn with the express purpose of proving they were liars and thieves. I can tell you that it's not a pleasant position to be in. All day I tried to get them to cooperate, while trying to prove: 'You guys are a bunch of cheats.'

"By the afternoon, the jet lag had really taken hold. I could barely stay awake. The guy in charge said, 'OK, you can come back tomorrow.' Next morning, he had changed his tune: 'Oh, no, no, you can't come back; no, sorry, you had your chance!'

"The lawyers had to go back in front of the judge, asking for more access. 'The poor guy is completely exhausted. It's not surprising since he's just come from England and gone straight to work.'

"I was granted more access. The more I looked at it, the more I realized they had attempted to rewrite the BBC MOS. They got a lot wrong. One thing I noticed was they completely copied the sideways ROM access mechanism, which is a weird piece of stack-juggling code to switch ROMs. I went back to my hotel room and wrote a completely different version of it, just to prove that Acorn's method wasn't the only solution, proving it wasn't a coincidence that theirs matched ours.

"They did some other dumb things they couldn't hide. They had copied the memory initialization routine – badly. We wrote a Return from Interrupt (RTI) instruction into the first byte of every 256-byte page. If we got any unexpected errors, it would just perform a return from interrupt, then keep running. They didn't realise what this was for, so they changed the memory initialization routine. Instead of stepping over this NMI that we put there, they wrote rubbish all over them. It was quite clear they had copied the MOS but didn't know what they were doing.

"The challenge was having to explain it in court, but we were lucky: we got the judge who had handled a copyright infringement case involving Apple, so he wasn't completely 'thick' when it came to computers; but despite that, this eminent legal mind was no computer scientist. We had to convince him – mostly based on the quality and strength of personality – that we were right, and they were wrong.

"While all that was going on, I visited a school where a prototype Squirrel had been installed. I discovered that although they'd rewritten the MOS (getting much of it wrong), they had just taken a BBC Basic ROM and put it in their machine. Legally, you can't use a competitor's product to make your own machine, even if you do plan to replace it eventually. When I presented my evidence, the judge put a complete hold on them, blocking the Squirrel. Costs were awarded to Barson Computers, but the defendant declared bankruptcy, so Barson carried the costs.

"I had some time to play tourist, so I bought a camera to record my exploits. I didn't realise how big Australia is! It takes four hours to fly from Perth to Sydney, which is a big city with some fabulous places like the harbour and the Opera House. I stayed at a place called Rushcutters Bay. The old part of the city was laid out in the late 1700s so it's not surprising that it reminded me of Central London – with small winding streets that change name regularly.

"I also had the opportunity to give the first public lecture on the ARM processor at Monash University near Melbourne, which is an august institution that felt a lot like being back at Cambridge. The presentation was well received. I enjoyed talking about what we were doing, its beginning, and how we made it work.

"Returning to Cambridge, I was reminded that the development costs of the Archimedes had strained Acorn's finances to the limit, aggravated by the poor performance of its Custom Systems division. Acorn's financial report of that year blames these for the company's losses. Acorn changed gear and moved away from specialist projects and focused on volume products. That meant 47 staff were made redundant, the Custom Systems division closed. Communicator development was abandoned.

"It had been a good time to be out of the country."

00011000

Kim Spence-Jones

Kim Spence-Jones graduated as a civil engineer and earned his MA in Engineering Science from the University of Cambridge in 1976. As a young graduate, he worked in the offshore oil industry where he developed a technique for connecting pipes underwater. In the late 1970s, with an interest in electronics, he developed industrial control equipment. This is how Kim remembers his time with Acorn…

"It was in 1979 that I first visited 4A Market Hill. I had produced an interface card that could connect an AIM-65 microcomputer to the Eurocard System video card. Acorn agreed to add it to their catalogue and paid me in kind with a System 2. I believe I got the better end of the deal because I don't think they ever sold any of my card.

"In 1980, I worked in Japan as the resident engineer supervising parts fabrication for a North Sea oil project. I took my System 2 with me and spent most of my spare time learning how it worked. On my return to Britain in March 1981, I dropped in to Market Hill. The BBC had just selected Acorn to build the BBC Microcomputer, so there was an enormous amount of activity. I was hired as a contractor to help develop the new machine.

"Four of us worked in a small office, packed in like sardines. I had a desk in one corner, Sophie had the desk to my left, Paul Bond and Laurence Hardwick shared the other half of the office.

"I was there when Fred, Jim and Sheila came into the world. FRED and JIM refer to memory addressing of the 1 MHz Extension Bus and paged RAM. SHEILA is the name of the resident hardware address space: video, cassette, sound and interrupts. We were in Chris Turner's office where he was amending the circuit diagram master copy on a drawing board that dominated the room. 'What shall I call these?' he asked.

"In honour of characters Fred Nurke and Count Jim Moriarty of *The Goon Show*, Sophie said, "Fred and Jim", so FRED and JIM they became. Chris added SHEILA later. He said the name just popped into his head.

"I'm often asked what it was like working with one of the greatest computer scientists of the 20th Century. Sophie is a true genius; and most people who have worked with her speak with respect and affection, tinged with fear and awe. She can be scary, sometimes a bit brusque. I don't think it's a conscious effort to intimidate others, but in a world where intelligence is the currency we trade, few of us would ever measure up. I hope I achieved *Not a Complete Idiot* status.

"Paul Bond, on the other hand, was more aware of other carbon life forms (including humans). As well as designing the first Machine Operating System, he ran interference between engineers and the rest of the company. He shared that role with Chris Turner, who also helped to keep the development team focused, and ensured that what we developed could be mass produced.

"The whole team (maybe a dozen of us) dined together most evenings – often as Hermann's guests – and we usually went back to work after dinner. Hardware and software problems were discussed, and solutions proposed during those meals. Though we visited a few different places, the *Eros* Greek restaurant was my favourite. Long gone now, it was in Petty Cury, just around the corner from our office in Market Hill. Main dishes were always served with rice and chips. That's what kept us fuelled up over those long summer days and even longer winter nights. Those dinners were essential in maintaining our rate of effort.

"Steve and Sophie had decided to cut down the number of discrete integrated circuits in the BBC Micro from more than 120 to something more manageable by packing some of the logic into Ferranti ULAs. I built a wire-wrap prototype of Chris Turner's cassette and RS-423 interface designs, as used in the Acorn Atom. This would become the Serial ULA.

"I spent long hours testing the frequency characteristics of different tape decks – from the cheapest models we could lay our hands on to my semi-professional Nakamichi. The output side was easy to design, once we realized that we could use blocks of the ULA to produce stepped approximations of the sine waves. I believe we mixed three separate component signals, then passed the result through a filter to create CASOUT, the ULA's cassette output signal. The input was a different matter. I spent weeks messing with component values in the filter circuit before I finally had something that worked well with all tape decks tested. The approach may seem unorthodox now; I just started with well-accepted theory and messed around until it worked. Years later, I learnt that Acorn had managed to source a BBC-labelled cassette deck that was of lesser quality than anything I'd tested. I had built an idiot-proof machine. Everything went fine until they found a bigger idiot!

"Once the ULA prototype was working, I drew the circuit schematic onto mylar film. It was passed to Mike Muller at the Computer Lab. (He later became a director of Arm.) He converted my drawing into a ULA design. Some years passed before I got to meet Mike in person. It says a lot about the pressure we were under to get development finished that we could work on the same project in the same town but never meet.

"Having cracked that nut, I got to design the analogue section of the video output. Years later, I was horrified to learn that components driving the composite output television signal varied from batch to batch. This forced manufacturing to develop a procedure for adjusting resistor values to compensate. Nobody thought to discuss this with me. It could have been fixed by changing the design to use the equivalent CMOS chip!

"Chris Turner and I shared the drawing of the original BBC Micro's circuit diagram – also on mylar film. Another part of my work was helping to lay out the main board. A professional layout draftsman, Bob Austin, did most of it, with me as his assistant, painstakingly checking every connection on the circuit diagram against the taped master. This involved laying black crepe tape and donut-shaped pads onto mylar film. The artwork was four-times actual size. These were photographically reduced to produce life-sized images that were used to produce the circuit boards.

"The process involved roughing out the design using red and blue pencils. Once that was checked, we replaced the lines with tape on two sheets of mylar. These represented the two sides of the printed circuit board. Having done that, and working only on a drawing board, we checked everything. Every change to the circuit involved replacing tape to reflect the new design. We often worked through the night, updating the layout with the latest changes.

"Once the BBC Micro was launched, I stayed at Acorn for a while, working with Andrew Gordon to develop the Teletext adaptor, which was based on guidance from BBC Research, and adapted by Mel Pullen. It wasn't my first choice for an assignment as I had no experience with radio frequency circuits. We couldn't get it to work. In the end, we asked the BBC Research guys at Kingswood Warren for help. They quickly diagnosed the problem, finding that one component needed a higher Q factor. Q describes the 'quality' of a tuned circuit or resonator. A higher Q means it is 'more inclined' to oscillate. It was a simple fix, once we understood the problem; and it helped to improve our relationship with the lads at BBC Research.

"I interfaced the first mouse to a BBC Micro and explored the design of a desktop bus codenamed the Pion. Like the Apple Desktop Bus (ADB) that connected the keyboard and mouse to early Macintosh computers, the Pion was to provide a common interface for all devices. But it never received the attention it needed and was finally dropped as a priority.

"Some time later, I returned to Acorn for one last project to help Brian Jones debug the wire wrap prototype of the Electron, at Fulbourn Road.

"In 1982, I established SJ Research to develop specialist networking products for the education market. Andrew Gordon joined me. We built Econet systems that performed better and operated faster over longer distances than Acorn's. Our secret sauce? We used higher-quality cable than what most installers used, along with our own terminators and clocks.

"The business focused on Econet, addressing market niches left un-served by Acorn, and where custom hardware could offer better performance than off-the-shelf products. Two examples were procuring a custom-made cable that precisely suited the requirements of Econet and building Econet interfaces for the IBM PC.

"SJ Research is also remembered for high-end file servers. There were three configurations of floppy disk and hard disk – the *Modular Disc File Server*. It was widely regarded as Britain's most reliable and best-performing Econet file server. I must agree.

"In 1996, with the end of Acorn Computers approaching, I built on my networking experience, forming SJ Consulting. Most projects involved design and implementation of advanced video and high-speed network systems. OpenHub was formed in 2004. The business implemented advanced digital systems that connected residential properties. More recently I've been working with central and local governments defining strategies for future smart homes and smart cities.

"People often ask me about Acorn's relationship with third-party suppliers like SJ Research. From a commercial perspective Acorn had no objection to any of its third-party suppliers making competitive products. There was no licensing deal involved. Acorn was always open to other people developing their own projects and supporting the Acorn ecosystem; believing that it all brought value to the product. We knew what everyone else was doing. I'm not aware of any jealousy; and there were always cordial relations between companies. Perhaps, in part, this was because most of us came from the same university backgrounds, so many of us had our start working for Hermann, Christopher and Andy."

00011001

Laurence Hardwick

Laurence Hardwick grew up in the Midlands and attended a comprehensive school where, like many bright pupils, he was singled out as being different. His way to escape bullying and isolation was to pass the entrance exam for the University of Cambridge. Here, Laurence describes his time at Acorn…

"I wanted to get into Cambridge because I'd had enough of how I was treated by others at school. I just wanted to be somewhere I might fit in and not stand out from everyone else. To meet the entrance requirements, I stayed on at school another year to pick up two more subjects: an extra English option and a foreign language. I chose German. With them, I was accepted to do electrical sciences at Trinity Hall.

"Forty years after going there, my hometown newspaper devoted four column-inches to say how fantastic it was that a chap from Nether Stowe Comprehensive School got into Cambridge. I don't know if others have achieved the same since, but I feel very lucky to have had that opportunity.

"One thing that helped me when I joined Acorn was that I had built a National Semiconductor SC/MP microcomputer while I was at university. The display used an unshielded television modulator that radiated so strongly that even students in the junior common room could tune in. It had no ROM or offline storage, so programs were keyed in, and it stayed on. As I had no printer, the instructions were written down in long-hand. At the end of term, the computer was switched off. Next term, I re-entered the operating system and got back to working on it.

"I liked the simplicity of the SC/MP. It would be a few years before Acorn started developing its Reduced Instruction Set Computer (RISC), but to me the SC/MP shared RISC features such as a simple instruction set and its memory addressing structure. Because of its simplicity, I could program binary codes straight into memory (without using an Assembler). It was a time when I could understand the entire design of a computer. I could conceive the structure – from the transistors and resistors to the gates and the overall processor design. I could understand how it worked in terms of the operating system and the interface to the TV and keyboard. Ten years later everything changed, computer development required a team of experts."

Cambridge Processor Unit

"I came to know Sophie Wilson through the university's Processor Group. She introduced me to Hermann Hauser. I started doing some work with Hermann and Christopher Curry while still studying for my degree. They were in their first offices in King's Parade, overlooking Kings College Chapel. There were two desks: If you sat at one you were working for Science of Cambridge; at the other, it was Cambridge Processor Unit (CPU). I never worked for Sinclair – only for CPU.

"When they moved to 4A Market Hill I did some work on Eurocards while still a student. My focus was on coding and analogue interfaces. Chris Turner tidied up Sophie's design of her *HAWK* computer, creating the Acorn Microcomputer and the System 1. As Chris was busy with other priorities, he had me check the filters of its cassette interface.

"When I graduated later in the year, I went on a 'milk round' – attending job interviews. One firm designing digital telephone exchanges was keen, but Hermann got in first. I was number six to join CPU, following Hermann, Christopher, Chris Turner, Sophie Wilson and Nicholas Toop.

"The Acorn System 1 was released through Acorn Computers. I was working for Chris Turner on System Eurocards, doing the track layout the old way – using tape on film. I didn't do much work on the Atom but some of my work for the System cards, like the cassette interface software, found its way into the Atom and the BBC Micro.

"Steve Furber was a regular visitor. Colin Priestley, Hugo Tyson, and others had joined, pushing the limits of the number of humans that the Market Hill ecosystem could practically sustain.

"For comic relief and to achieve a little celebrity status, I published a *Bulls and Cows* game – what is now called *Mastermind.* I also published a chess program. They could fit in the 1.5k byte memory of the System 1. They're available online in the *Liverpool Software Gazette*, November 1979. In the same edition Acorn promised to release a 4K integer BASIC in EPROM by January, with floating point in the works, and claimed that a Motorola 6809 Eurocard was in wire wrap. They were fun times."

Econet

"Possibly my most important contribution at Acorn was the Econet interface, which began for the Atom soon after joining the company. It started as ideas proposed by Andy Hopper. Networking was to become a big thing – bigger than many people realized at the time. I later became Econet National Product Manager and an enthusiastic advocate for it.

"Hermann and Andy Hopper set up Orbis to develop a commercial version of the Cambridge Ring. Everyone at Market Hill was aware of it. We talked about how to get microcomputers to talk to one another too – what would come to be called a local area network. Andy had the idea of using an IBM communication standard called High-Level Data Link Control (HDLC) that became popular during the 1980s. The chips were reliable and cheap, so an Eco-nomical net-work for the Eurocard Systems and later the Atom seemed viable. Thus, the *Eco-net* was born.

"Today there are many ways to connect computers together for sharing data; back then the concept was quite new. Andy suggested a simple method where all machines share common connecting wires using a technique called Carrier Sense Multiple Access/Collision Detect (CSMA/CD). Collision detection and collision arbitration simply mean that multiple computers access a shared line with a common carrier signal. Each computer checks before transmitting to ensure the line is clear. If a collision is detected while attempting to transmit, the affected computers wait for random time delays, then try again; repeating these steps until the transmission is sent. This all happens in a fraction of a second, so even if there is a collision, it is resolved without users being aware of it.

"Hermann passed me ideas that Andy had drawn up. He proposed using a Motorola 68B54 chip that could easily support up to 256 devices. After doing some calculations, I found it needed some changes in the way it sensed the signals on a CSMA/CD network. Andy's original design also needed components adjusted during manufacture, making it impractical for volume production.

"Applying what I'd learnt when studying wave guides at university, I decided to place a voltage divider across a differential pair of signals. This is where one conductor goes positive while the other goes negative to represent a logic 1, and the conductors reverse their polarities (reverse the positive and negative voltages) for a logic 0. I proposed using a comparator to determine if the pair was above or below a set voltage level to detect logic 1 or 0, or a collision.

"I calculated how fast signals could travel along the cable. At first, Econet didn't need to run at the maximum possible speed. I chose a clock with an equal mark-to-space ratio. This means the clock signal spends equal periods of time in its two states: high and low. This worked fine with the System and Atom machines, which were slower processing the data. Later, Kim Spence-Jones came up with the asymmetrical clock. Changing the mark-to-space ratio allowed the Econet to run a lot faster, delivering higher speeds which the BBC Master needed.

"Networks like AppleTalk reconstruct the clock from the data. I chose a separate, dedicated clock to synchronise the data signals. The interface circuit is much simpler and very reliable. Market Hill already had a lot of unused four-core telephone wire installed between floors, so I connected extra cable to prove I could achieve maximum cable lengths.

"When wires are bound together, external noise interference appears on all of them at roughly the same time – common mode signal. Twisting two wires around one another (a twisted pair) cancels out a lot of that noise. The incoming signals pass through integrated circuits called voltage comparators that eliminate most of that noise. Econet proved its ability to block common mode noise when the only network that the Lydney iron foundry could run was Econet, after unsuccessfully trying Ethernet and other technologies.

"In later years, some played at the edges of the original Econet design. Other than Kim Spence Jones' work on the clock signal, most attempts didn't end well. There was a time after I moved to Australia that someone in Cambridge tried to reduce production costs on the Econet interface (a waste of effort because it was already cheap to make). I received a call from Cambridge asking me, 'What happens if we remove the collision detect? Since we're already handling errors, does it really matter?'

"I objected, but my protests fell on deaf ears. Some time passed, then I received another call. 'Why isn't this working?' It seemed perfectly clear to me: 'It's not working because you've taken out the collision detection!' They replaced it, and sure enough, everything was fine again."

Cassette Filing System

"Applying what I learnt with the Eurocard System cassette interface, I wrote the Cassette Filing System (CFS) for the Atom. I had to count every processor cycle when generating sine waves to ensure the output was as smooth and pure as possible. To get a uniform data stream every clock cycle mattered. The output coming from the tape was as perfect as I could make it in eight bits of data.

"I also developed the CFS sequential file access method, which split the data stream into 256-byte blocks, with a break between each, to keep count of which block was being accessed. Using a tape recorder that can fast-forward and rewind, a BBC Micro can find the start of any block, then rewrite a new block into the middle of that part of the tape – an addressable cassette filing system. This also helped when loading large programs as most tape read errors could be rectified by just rewinding to the start of the misread block and loading it again. This 256-byte data block size was retained when the file system was updated for the BBC Microcomputer."

Supporting a million machines

"Following the announcement of the BBC computer contract Acorn's phone lines were running white hot, so I started doing technical support. By the time we moved to the Waterworks, I reported to Sandy Dow while overseeing all support. Even before we started shipping the new design, we had a team who were dealing with the constant demands from people who hadn't even seen a BBC Micro. When we started shipping machines, we answered lots of phone calls and letters asking: 'Where's my BBC computer?' As well as leading this team I acted as roving technical support for computer exhibitions, visits, and major events.

"Sometimes getting machines prepared for events called for a little ingenuity. Everything gets hot under heavy lights at trade shows – especially the computers. The early BBC Micro linear power supplies were susceptible to overheating. This added to video ULA problems where overheating caused computer displays to fail. To keep machines cool, I bypassed the on-board power supplies and connected the DC inputs to large external supplies, using the mains power chord to disguise what I had done. It all seemed normal to a casual observer.

"At a *PC World* show in London, which was one of the first public appearances of the BBC Micro, John Coll and I were the floor show. Because it was shoulder to shoulder, I remember John and I spent each day standing on top of the exhibition stand with our heads through the roof, giving forth to the crowd who were eager to get a glimpse of the new computer. I got to know John quite well, and I must say there was a lot to like about his quiet confidence and forthright approach to bringing people along to his way of seeing the future.

"When we moved to Fulbourn Road late in 1981, Hermann bought me a company motorbike as I couldn't drive a car at the time. I travelled around the country on it promoting networking in schools. My enthusiasm for the Econet made up for my lack of training in sales and marketing.

"In 1982, I rode the company motorbike to Kings Cross where I placed my helmet and coveralls in a locker at the station; then I attended an event on the *IT82* Train. It was traveling around Great Britain as part of a public education programme showing off the products of Acorn, Sinclair, Research Machines, Newbury Laboratories, and other British computer manufacturers. I'd been doing demonstrations and learnt what other companies had to say – most of it word-for-word.

"The event was a visit to the train by the future King Charles III. When he and his retinue boarded, there wasn't much room, so we all played musical chairs, shuffling around in the carriage to make room for the visitors. When the music stopped, I found myself standing in front of the Research Machines RML 380Z. Prince Charles stepped up to me and asked, 'What does this one do?' So, I demonstrated the RM computer following the script that the Research Machines guy had been using all day. He was at the other end of the coach, unable to get to his machine. He gave me a thumbs-up, mouthing: 'You're doing a great job!'

"In 1983 I was sent to India with a civil servant called Morris. He was there to fly the technology flag for Britain, and I was to look after computer networking and explain it to the Queen and the Duke of Edinburgh. This was during the *Seventh Commonwealth Heads of Government Meeting* in New Delhi. The Queen was going to present six networks, each with six BBC Micros to the Indian Department of Education. The Queen and Prince Philip needed to be briefed before they handed over the computers, just so they had some idea of what they were presenting.

"We set up the networks in the hall of an Indian Army barracks where the presentation was to take place. I walked along one bank of computers talking to Prince Philip, while Morris was speaking to the Queen.

"The accommodation was awful because all the decent hotels had been taken by conference delegates; but we did get out to see a bit of Delhi. I remember they'd painted white lines down the middle of the roads but none of the taxi drivers knew what they were for. It was still chaos.

"As the company started to focus on the possibility of developing a US presence, I made a few trips there with John Coll. John Caswell designed knockdown stands and shipped them to the United States where John Coll and I assembled them. We toured the USA, demonstrating the BBC Micro in support of promotions for the *Computer Literacy Project.* We had a once-in-a-lifetime opportunity to visit *Sesame Street.* We also visited MIT to meet Seymour Papert. He enjoyed superstar status as inventor of the Logo programming language and the robot turtle.

"Customer Service became one of Acorn's largest divisions. David Bell was recruited early in 1983 as a project manager, and to liaise with the BBC. Around that time, Mike Bicknell was appointed as Customer

Support Manager to direct our expanding support team. That was a load off my mind."

Continuing Econet development

"I later became Product Manager responsible for all Econet development including the Econet FileStore. There were a few in management who didn't understand or support Econet in the way they should. They ignored the fact that we were selling 20–30 BBC Micros at a time into schools simply because they could link them up, share one disk, and run class sessions. I sometimes had to fight to get Econet products manufactured for stock; but our efforts allowed machines with the Econet interface to sell in the greatest numbers around the world. We dominated that market until Apple caught up.

"I felt I needed a change after that, and told my manager that I had to find something different to do or move on.

"Sam Wauchope our MD at the time had decided to take direct control of the distribution of Acorn Computers in Australia. He and my manager David Bell decided I would be a good fit to join the Australian team.

"I would have loved to stay as Australian National Product Manager, but after seven years Acorn had one last task for me: I returned to the UK to take up the role of Licensee Support Manager for the Acorn Network Computer (NC). Again, I was travelling the world, but this time working alongside engineers and marketing teams from Oracle promoting the NC concept.

"I'd spent 18 years working for Acorn in different roles, seeing many of its ups and downs. On reflection, I enjoyed those times and worked with some great people, but it was time to move on."

After Acorn

"Econet has a lot in common with other modern communication standards such as the CAN bus, which is used to connect equipment in vehicles, boats and aircraft. Having it on my resumé led me to work on two projects with Malcolm Bird. (It was Malcolm, in his capacity as Acorn's latter-day Chief Technology Officer, who dragged me kicking and screaming back from Australia to promote the Oracle NC.) I also worked with Malcolm as Chief Engineer for Europe, the Middle East and Africa for WAP Technology developed by Unwired Planet, designing WAP data centres.

"Malcolm called on me again to work for a company that built composite fibre ultralight aircraft. It used CAN bus to integrate engine controls with the avionics and strain gauges, so I read a lot of raw CAN bus data and got quite good at translating and analysing its messages."

On Retro Computing

"I think the attraction to retro computing is its simplicity and accessibility. I still have a hobbyist interest myself, but I use Raspberry Pi. I do different things with them and publish my code on GitHub for others to use. I write in Python, JavaScript and some raw HTML. My Python programming starts out looking like well-structured BBC BASIC, but then I rewrite it as more modern object-oriented code.

"I still like to solve engineering problems."

00011010

Paul Fellows

Paul Fellows joined Acornsoft toward the end of 1981 while still at university. He became one of its most valued and longest-serving software specialists. This is Paul's story…

"Acornsoft was incorporated to develop software for the Acorn Atom, soon after the first machine was shipped in 1980, with David Johnson-Davies as Managing Director. Its focus was on games software but it also developed programming languages and some business applications.

"On my first day at Market Hill, Acornsoft had six employees: David, his secretary Nikki, Phillipa in charge of product distribution, Chris Jordan as publications editor, with Tim Dobson and Jonathon Griffiths writing games. The rest of us were independent contractors. We were paid royalties or outright for the software we delivered. I wrote a few educational programs; the first was for chemistry. I also wrote a game called *Sphinx Adventure* and then wrote *Database*.

"The company provided a channel for external software developers to sell under the Acorn banner. Its mix of third-party developers, contractors and employees followed the same approach as the parent company. This helped to meet consumer demand for software, while spreading development costs during its startup phase.

"David first worked with Christopher Curry in 1977 when he wrote *Moon Lander* for the Science of Cambridge MK14 computer. It shows the rapid advances in games software over those three or four years – even when comparing it to very basic computers like the Acorn Atom. On the MK14, *Moon Lander* displayed the LEM's speed, height, and fuel consumption on an eight-character calculator-style display, whereas Atom video games such as *Space Invaders* used colour displays, animation and sound, along with more interactive user inputs.

"David later wrote the manual for the Acorn Atom – *Atomic Theory and Practice*. It reflected the quality and attention to detail that became a hallmark of Acornsoft products. He was joined early in 1981 by Tim Dobson and Chris Jordan – as the BBC Microcomputer project was getting underway at 4A Market Hill.

"As Acornsoft grew in experience and confidence, it published original ground-breaking games for the BBC Micro, such as *Elite*, one of the first intergalactic trading games, and *Aviator*. It also produced one of the world's first racing games *Revs* – a Formula 3 simulator. Some were remakes of popular arcade games: *Hopper* is a clone of Sega's *Frogger*. *Snapper* and *Arcadians* are based on Namco's *Pac-Man* and *Galaxian*.

"The mainframe computer system at the University of Cambridge was fertile ground for finding software developers. Mathematicians Jonathan Mestel and Peter Killworth wrote groundbreaking text adventure games which Acornsoft released. To this day, interactive fiction enthusiasts believe *Philosopher's Quest* and *Countdown to Doom* are among the best text adventure games ever produced.

"Peter was a keen amateur magician. He wanted to show that computer games didn't need complex graphics or animation to be entertaining. In 1984, Acornsoft released *The Paul Daniels' Magic Show* – 10 magic tricks, games and illusions, designed to entertain groups of players around a computer by following on-screen instructions. While they would be considered a bit twee today, this computer game was very popular in the 1980s. He later worked on the RISC OS graph plotting program, *Tau*.

"Acornsoft quickly became the UK's leading publisher of high-quality computer games. It also published versions of popular ZX Spectrum games on Acorn platforms and some notable education packages, equal to the best from the Microelectronic Education Programme, BBC Software and other educational software publishers. As well as their games catalogue, they were well known for publishing programming language compilers and business software.

"Successful developers have to deal with software piracy; Acornsoft was no exception. In 1984, the High Court granted Acorn Computers an injunction requiring Computer Publications to withdraw all copies of its January edition of *Personal Computer World*, which described how to bypass Acornsoft's copy protection. The Court found that PCW had incited readers to pirate computer programs. This cost Computer Publications £65,000 plus costs in an out-of-court settlement, and the cost of reprinting the magazine, estimated to be another £100,000.

"After completing my first degree in chemistry, I decided to study computer science at the University Computer Laboratory. I wrote a Pascal compiler for the BBC Micro as my major project. David bought the publishing rights and offered me a job running the computer languages group. Till then, Jeremy Bennett had been on his own. They hired me and four other guys because programming languages were in great demand.

"We published several languages and developed their test and demo programs. There was Forth, Lisp, BCPL, Pascal, Comal, Prolog, and of course there was Microtext. It was a bad decision to foist that project on us. David didn't want to do it, but it's an example of how things were decided above his head. Christopher Curry had agreed to publish Microtext for the National Physical Laboratory in London, even though it was expensive to produce. In my opinion, an enormous amount of time, effort and money was wasted. It was just one example where others within Acorn held sway over what we had to produce.

"Our lineup included David directing the Games team with Tim Dobson, Jonathan Griffiths and some external authors including Neil Reyne. I looked after the Languages team with Stuart Swales, Tony Thompson and Richard Manby. Our job was to test and report bugs in the software, write demonstration programs, and get the manuals ready.

"Phillipa Bush ran a small team in publications, which included my wife Sharron, who wrote quite a lot of the manuals. Sadly, she passed away in 2011. Shelly was looking after marketing, along with Ellis Hall. There was also Alan Bellingham and two other chaps in the tech services team doing testing and IT work.

"Rob McMillan, also from my college, joined Acornsoft to take charge of the business software. He worked with Paul Hudson in-house, and Mark Colton who was the author of *View*, *View Chart* and *View Sheet*. It was the same business model that I had with software developers like Arthur Norman (LISP and Chess), Richard DeGrandis-Harrison (Forth) and Martin Richards (BCPL).

"*View* was a great word processor, but boy did it take a long time to develop! Computer Concepts had released *WordsWise* well before it, so they captured most of the market. The time to get *View* published reflects some of Acorn's internal growing pains after its early successes. Mark Colton did a great job getting it right; but getting it produced and released took a year, which was far too long. It was one of Acornsoft's first ROM-based software packages. Till then, everything was published on cassette tape or disk. Having to go through the Acorn mothership, we became bogged down in funding approvals and their process for making ROMs.

"Getting 6000 masked ROMs made was expensive at the time. It was certainly beyond Acornsoft's budget, so it took time to convince Acorn to spend the money. In hindsight, they were right. *View* sold for about £80–99 in the UK, so we would never have made a profit selling it. We had to give 50% margin to resellers like Watford Electronics. Starting at

£99, then taking 20% off for VAT, we probably got £80. £40 went to the reseller. Of the £40 left, we paid Mark Colton £8. The ROM, box and manual swallowed up what was left so we made virtually nothing for the effort and the risk we were taking.

"It's all very well talking about an Acorn ecosystem and encouraging other people to get involved, but I believe everything should be profitable. Acorn never seemed to be commercially savvy. I would often ask why we made some of the decisions we did – particularly with pricing. Acorn, for example, sold fewer disk drives than it could have because others blew copies of the DFS and got all the bits together to make their own. They duplicated the disks and copied the manuals. Acorn wasn't making anything out of that. That happened with Econet too, and it happened with other upgrades and add-ons.

"Fifty per cent discount on the software never made sense. We were ruled by the retailers. Imagine trying to do that to Apple! People like Watford Electronics were keen to sell Acornsoft products because of the generous discounts. Did the Commercial Manager ever question why we were turning over so much stock, but just breaking even, at best?

"I don't know who set the prices, but they were guaranteed to lose money. *Elite* is another good example: It went out the door for quite a high price I think it was £18. We had about £8 coming in. But take away the royalties, disks and packaging and I believe we were left with precisely zero pounds to show for our trouble.

"Before Olivetti took over in 1985, we just did what we were told, and nobody calculated the cost. I could see things happening that didn't make sense. I did try to figure out what was going on, but we were all 'discouraged' from getting involved. After the takeover, it took my new boss no time to discover that we were losing money. He was astonished that we didn't have a business plan, a costing spreadsheet, or anything to show how things were going; and how we were going to make money. The first thing he asked me to do was to work out what everything cost us. It was an absolute shock to find that our software was going onto the market at prices set below cost. It was clear that Acornsoft was haemorrhaging cash.

"How could they not make money in that business? As far as I can tell, they never tracked the costs. They didn't ask: How many are we going to sell? What price should be charged? How much did things cost? Little things escaped us like the fact that we were paying two pounds for a printed wallet and a couple of pounds to get the disks duplicated, then a pound or two in royalties to the author. They were costing us a fiver, and we were giving the retailer a 50% discount on a program selling for £9.99. There was nothing left.

"On top of the cost of sales, there were the overheads. We employed people to test the software and write the user manual; to store, pack and ship the product to the reseller; and there was also the cost of keeping the office open. It would have been different if we sold directly. We would have kept that 50% instead of giving the margin away to the dealers who argued that they needed it because they put ads in magazines.

"There was an attitude that Acornsoft existed to support sales of the computer, so it didn't have to make money. 'We are a loss leader.' We were reminded whenever we questioned the pricing. We were told not to worry about it; but when my boss from Olivetti asked me, 'What are you doing? Why don't you know whether you're making money?' I had to explain that we weren't allowed to know pricing details.

"I was shocked when the accounts department went through it with me. We couldn't sell any more games because we couldn't make any money out of them. We could make money on the languages because there was some wriggle room – provided we kept costs under control.

"Once we became involved in setting the prices, we developed a few more packages: the BASIC Editor, the Graphic Extension ROM, and the C and Comal programming languages. We had to increase the prices and cut the retail discount to a much more reasonable 25%.

"When Olivetti finished investigating the financial position, this marked Acornsoft's end as a developer. As long as Acorn Computers was profitable, Acornsoft might have been able to exist as a loss leader, though better pricing arrangements would have been a smarter business decision. But it was too late to save the business from the position it was in. David Johnson-Davies resigned.

"A year later, most of the Acornsoft games catalogue was sold to Superior Software. The text adventure games were sold to Topologika, which updated and re-released them on other platforms.

"Office software – *View*, *View Sheet* and *View Base* – continued under the Acornsoft brand for the BBC Master series. We all moved to the Acorn site at Cherry Hinton, where we were nominally still working for Acornsoft, but our pay cheques were coming directly from Acorn. That lasted for about a year.

"In September 1985, I was called into what looked like a high-power meeting. Among those present were Sophie, Jim Merriman (Technical Director) and Hermann. I didn't sit down; I just stood there. It seemed the right thing to do.

"Well-known for his directness, Jim asked me, 'What are you guys doing at the moment?'

"I explained that we were finishing off the C compiler for the BBC Micro and the 6502 second processor. He followed in rapid fire:

'Right, how would you like to write an operating system for the first machine to run the ARM processor?

'Can you drop everything and do that?

'You've got five months to do it before the hardware is ready.

'We want a BBC Micro-like operating system cloned for the ARM.

'Can you do it?'

"I said, 'Yes, we can.'

"Jim had a way of dominating conversations. Finally, impatient to get things moving, he said, 'Ok then, go and get on with it.'

"That's how *Team Arthur* was born with Stuart Swales, Tony Thompson, Richard Manby, Tim Dobson, Neil Reyne, Nick Reeves and me.

"I believed we could do it because my group had written the Graphics Extension ROM for the BBC Micro, which was then absorbed into the BBC Master's operating system. We knew about some of the more obscure aspects of the operating system, having developed sideways language ROMs and utilities. We knew the finer details of how to produce graphics, the user interface, file systems, and everything else that would be needed for the Archimedes.

"We understood some of the ARM architecture. Some weeks earlier, when we asked for an ARM second processor, they told us we could have the parts, but no one was available to build it. I sent my guys down to the store where they managed to scrounge what was needed. They signed out the spare parts for two ARM second processors. We built them ourselves, played with them, and decided: These are fantastic!"

The Culture Shift

"Acorn's workplace culture had been formed in the late 1970s and early 1980s by people like Hermann, Christopher, Chris Turner, Steve Furber and Sophie. They were also the mainspring of the Systems, Atom and BBC Micro successes. People were hired along the way, often through introductions from Andy Hopper. Acorn had been selling computers as fast as it could make them.

"Once Acorn went public, the culture of the company changed almost overnight. They hired people like Jeff Tansley, Andrew McKernan and

Mark Jenkin who came from the 'respectable' side of the computer industry. They knew about VAXs and PDP11s from Digital, and mainframes from ICL and IBM. The new managers grabbed the reins of projects and were quick to tear down things we had spent years building.

"They all came from workstation backgrounds. Their view of the future needed a lot of money to develop machines in a market where the USA already had a commanding lead. They created a split within the company, and it really struggled to find its direction. In my opinion, that's what destroyed the company. They failed to understand that the beating heart of Acorn was trust, teamwork and the way people were empowered to do the impossible with precious few resources – that's what had made Acorn so successful.

"This wasn't the first time there was a difference of opinion about what sort of computers Acorn should build. That's why Sophie came up with the concept of the second processor. She talked about the days before the BBC Micro in 1979–1980. The lively debates between her, Steve, Christopher, Herman and Andy Hopper. Andy wanted to build workstations. Herman and Christopher wanted to build BBC Micros. Steve and Sophie just wanted to build things they could be proud of.

"The new guard took charge of the NS32016 project, the early Cambridge Workstations, the Acorn Business Computer, the Z80 Second Processor with CPM on it, and the Communicator. All of that was managed by a group of guys about 10 years older than the rest of us who held the view that the BBC Microcomputer wasn't a 'proper' computer. It was a toy. Acorn should be making proper computers (whatever that meant). They consumed vast amounts of resources and achieved little return on investment. Trying to turn Acorn into a 'proper computer company', they wasted millions of pounds in the process.

"It was a similar problem with the Acorn Research Centre in Palo Alto: It spent too much money creating an ARX operating system for a workstation idyll. For all that effort and all that money, what did they have to show for it? ARX never saw the light of day.

"Working on Arthur insulated our team from most of what was happening. It wasn't a conscious effort, but we effectively destroyed ARX from the inside by developing Arthur. Sophie never believed in ARX. Her long game was to have Arthur as a stopgap measure to buy her more time so she could develop an operating system more in keeping with what Apple had developed for the Macintosh. She sees Arthur and RISC OS as quite different beasts, but I disagree. It was an evolution from one to the other, with Arthur providing the foundation of what followed.

"My line manager was David Bell – the BBC Micro Product Marketing Manager. He looked after our 'pastoral care', ensuring we got paid at the end of the month. Sophie advised me on technical matters, but we seldom spoke more than once a week. With so many other priorities at the time, she was happy as long as things kept moving forward and was informed about what we were doing. In effect, she left me in charge of it.

"We started with a second processor board that just had the CPU plugged into it. Then we got a new second processor that had the I/O controller (IOC) and video controller (VIDC) chips on the main board – a so-called A500 second processor unit. We made good progress with that. Within three months, we went from a standing start to getting the video and keyboard working; then we received an Archimedes A500 computer. Pretty soon, we had one for each of us.

"After loading the embryo operating system onto the A500 machine, we were able to unplug the Tube and get it working and rebooting on its own. After five months, we had the disk drives, sound, and everything else working. Functionally, Arthur was complete, but it still had bugs. It took another four months to debug it completely. The hardware arrived late, so five months had been extended to nine. When machines were finally ready to ship, so was Arthur.

"About halfway through the project, Neil Reyne became available and joined the group. He wrote the Font Manager, which he did very nicely and very quickly. Till then, all text was hard-edged on computer screens. Arthur predates the Macintosh in displaying anti-aliased text – a relatively new technique at the time, then Neil wrote the Window Manager and got it working in four weeks – what an incredibly clever bloke he is!

"Everything was written in assembler. There was a C compiler, but it only worked with ARX. The workstation team were writing it in the other building, blinded by their workstation ethos. They absolutely would not cooperate with us. Their attitude was: 'Why are you wasting your time building that toy? That's not a real operating system. Go away!' That tells you a lot about a company in decline, because that's the antithesis of what Acorn was like in the beginning.

"Arthur had started as a small group but over time others joined us as they saw what we were achieving. Brian Cockburn wrote the Econet software. As more parts of the system began to work, more people began to see they should get behind it. A crucial moment came when we heard there was going to be a board meeting. They told both groups to present what they had achieved: ARX and Arthur.

"We learnt that the ARX group was going to run a clock program in a window, then run more copies to show it multitasking – their message being that theirs was a proper operating system; the implication being that Arthur would never be able to do any of that. So, Tim Dobson wrote a clock program that ran under Neil's Window Manager. It could run lots of them. We didn't tell anyone before we rolled up to the meeting.

"The ARX guys went first and proudly ran the demo of their clock program. In one window, it worked smoothly, the second one too; but when they ran the third window, the hands started jumping in two-second steps. By the time they had five running, it was lumpy, because on a four-megabyte machine with hard drive, it swapped virtual memory.

"We turned up with one of the first real production machines that had only one megabyte of RAM. It ran 19 copies of the clock program in separate windows, all scrolling smoothly. With that, people at Acorn threw in behind Arthur. They didn't immediately abandon ARX, but it didn't last much longer. This all happened around July 1986.

"There had been a rush to produce software during the 1980s on the back of the Acorn Atom, and later the BBC Microcomputer, Acorn Electron and BBC Master. The limited memory capacity of 8-bit machines meant that programs had to be written in machine code. That attracted a higher standard of software developer. This fertile field of talented programmers formed the core team of system programmers at Acorn. This is how such a small company could create an operating system with advanced capabilities that ultimately became RISC OS.

"From 1987, Acornsoft reappeared, actively promoting and supporting affordable RISC OS-compatible software for Archimedes.

"It was my time to move on. Arthur would evolve to become RISC OS. Tension between camps in the silver building was counterproductive. It needed someone who could rebuild bridges. I argued that development should be led by someone who had the personality to bring everyone together. William Stoye had proved his worth in the Workstation group. Bill replaced me leading the same team that developed Arthur, though other excellent people joined later. Together, they refined and extended RISC OS, which continued to be supported, years after the hardware was removed from sale."

Note. Development of RISC OS draws different views from Sophie Wilson, Paul Fellows, Carl Dellar and David Bell. While reviewing their differences, I asked Bill Stoye for his view. He sees Arthur as the first-generation foundation on which RISC OS was built. The main window, context sensitive menus and task bar were features of Arthur 1.20. There never was Arthur 2.0 or RISC OS 1.0. (It's said that the name changed to avoid confusion with the movie *Arthur 2: On the Rocks.*) RISC OS 2.0 extended Arthur by adding cooperative multitasking, multiple windows and anti-aliased fonts. Draw, Edit and Paint applications are like MacDraw, MacWrite and MacPaint on the early Macintosh.

00011011

Ramanuj Banerjee

Wire wrapping is a prototyping technique used in modern electronics to electrically join IC socket pins together. The connections are made by wrapping one end of a piece of fine insulated wire around a pin, then wrapping the other end around a different pin.

Ramanuj Banerjee, better known as Ram, achieved what seemed to be the impossible in a 35-hour non-stop wire wrapping marathon. Under Steve Furber's direction, he successfully prototyped the Acorn Proton. This enabled Steve and Sophie Wilson to demonstrate a working machine when the BBC came to call the following day. Without their teamwork and commitment, it's probable that the contract would have been awarded to a different company and the history of Acorn, ARM and so many other successful UK businesses might never have been.

On that first Monday in February 1981, Hermann Hauser phoned Ram in urgent need of his help. He explained that the BBC was coming in four days to see a working Acorn Proton. At the time, Steve and Sophie were huddled over Chris Turner's drawing board, having borrowed his office to get the space and resources they needed. They were busy pulling together some partly developed ideas for a new microcomputer.

The stakes were high – though no one at the time knew just how high. Winning the contract to build the BBC Microcomputer under licence would be a big leap forward in the fortunes of Acorn Computers, its shareholders, and those who worked with them. Hermann often demonstrated his ability to persuade people to achieve the impossible. Asking Ram to build the prototype in a couple of days was one of those times. Here is Ram's story…

"I promised Hermann to drop most of my plans for the week but couldn't come until the following evening due to my study commitments at the University Computer Laboratory. I arrived at Market Hill the following evening at six o'clock. After a short briefing from Steve, we set to work.

"A wire wrapped circuit prototype looks like multi-coloured spaghetti. Bugs are usually due to one of three faults: a connection to the wrong pin;

terminations touching each other when there should be space between them; and occasionally, a broken, loose, or unterminated wire. To reduce the chance of a wire being connected to the wrong pin, and to simplify debugging, I used colour coded wire; with a trick to choose the colour. This is how I did it:

"The numbers of the start and end pins are added together, then a 'modulo 10' addition is made. That simply means remove all digits that have a value of 10 or above, just leaving the units. The result, which could be any number between zero and nine, determines the wire's colour. It follows the same numbering system used for resistor colour codes:

black=0, brown=1, red=2, orange=3, yellow=4, green=5, blue=6, violet=7, grey=8, white=9.

So, for example, to connect pin 3 on one chip to pin 15 on another:

(3 + 15) mod 10 = 18 mod 10 = 8

Referring to the list of colour codes, the wire would have grey insulation.

"Here is another example. Connect pin 14 on one chip to pin 27 on another:

(14 + 27) mod 10 = 41 mod 10 = 1

Again, referring to the list of colour codes, the wire would have brown insulation.

"There were more than 120 integrated circuits in the prototype's design. A battery-powered wire wrap tool helped to make the 2,400 connections. Precision was vital. A wiring error can take an hour to locate. I had to do the job quickly and accurately. Repetitive tasks like this can be prone to error. I had the right tools, techniques and support from Steve and others doing the error checking. That made the difference.

"We had a couple of radiators burning away the whole time we were working. The Market Hill building was always cold; and being the middle of winter, it was freezing. Hermann, Christopher Curry, Chris Turner and others popped by from time to time bringing food, hot cups of tea, and moral support. There was always someone there to call out the pin numbers and calculate the colour of the wire to use. Steve and Sophie had to make design changes on-the-fly, which added some uncertainty, so the changes all needed to be reviewed during the debugging.

"To minimise the chance of errors, we checked and re-checked our work, testing the continuity of connections and the isolation between neighbouring pins as we completed each section. A few bugs did slip through. When I left at five o'clock on Thursday morning, all sockets powered up. Steve, Sophie and Hermann could start populating sections of the prototype and progressively test each one.

"Finding my bed where I'd left it a couple of days earlier, I collapsed in a heap. Fatigue doesn't encourage a deep and refreshing sleep. I awoke a few times during that Thursday. I would have liked to check in on Market Hill but felt I couldn't move without doing myself an injury. I finally looked in on them at about eight o'clock in the evening. The machine still wasn't working but Chris Turner's office was already overcrowded, so I went home to get more sleep.

"I didn't hear from them until a week later, when Hermann called me at the Computer Lab. I had just returned from morning tea and had brought mine back to my desk. He told me that it had taken all day and all night to get the prototype working. The BBC people came on Friday morning and were happy with what they saw. On the following Thursday, they decided to award Acorn the contract to build the BBC Micro. I was so excited when he broke the news, I spilt my tea over the papers I was working on. I had to go to Andy Hopper and ask for another copy. He didn't mind. He was equally excited. Hermann had called him too.

"Though I was very happy with what I received as a generous hourly fee, Hermann insisted that he wanted to give me something extra for my effort. Despite my protests and seeing that I was not about to get my way, I agreed to accept a thank-you gift. I asked whether I could have a computer for my younger brother. He told me to go to the store and take whatever I wanted. So, my brother Bobby received an Acorn Atom. It proved to be a good choice. Some years later, Dr Bobby Banerjee received his PhD in Artificial Intelligence from the University of Bristol.

"I completed my studies in 1982. After the summer break, I joined the Acorn payroll in September. I was working with Mike Muller and Andy Hopper installing Cambridge Rings at the University. We were operating as Orbis at first, but it was later absorbed into Acorn. Eventually, Cambridge Ring was dropped as a product line. Mike and I worked in the Advanced R&D group, investigating ways to use and interface the latest products – things like high density storage and laser printers.

"Fast forward to late November 1982 at Las Vegas Airport. As I passed row-upon-row of one-armed bandits (even in the arrivals hall), I wanted to say, 'Beam me up, Scotty; there's no sign of intelligent life here.' But since my teleporter was in the shop for repairs, I was stuck here at COMDEX/Fall '82 – one of the world's biggest computer exhibitions.

"This was the show where Bill Gates first saw VisiCorp's Visi On – the first GUI software suite for the IBM PC and compatibles. It's claimed to have sparked development of MS Windows 1.0 and other window-based operating systems; and may have influenced Arthur and RISC OS.

"I was due to be married in Calcutta on the tenth of December. Despite all my protests, Hermann had talked me into coming to COMDEX. Because I designed the Motorola 68020 second processor, he wanted me there, just in case. He said, 'Whatever happens, I'll get you to your wedding on time.'

"Acorn didn't have a stand. Hermann hoped to convince another exhibitor to allow us use of some of their space. In hindsight, attending COMDEX that year was hardly worth the cost. Sometimes taking risks pays off; and sometimes it doesn't. Being young and impressionable, I was amazed by the excesses of – well – everything. I was particularly taken with the all-you-can-eat buffet. America is a country, where doing things to excess never seems to be enough. On the fifth of December, I flew back to the UK after saying farewell to the gambling, the alcohol, and the people looking like they had lost all sense of reality.

"I returned to London Heathrow Airport the following day, ready to board my flight to Calcutta, only to learn that the departure time for my flight had been moved forward by three or four hours. The chap at the British Airways check-in desk was unsympathetic, saying, 'The travel agency should have advised you of the change.' I replied, 'Well yes, they should have but the fact remains that I am getting married in five days' time, so we need to fix this.'

"I tried to tell myself, 'Don't worry, I have three days to get there.' Then, I thought, 'My father will kill me!'

"I pleaded with him, 'Please, you must get me on another flight.' He said, 'Nothing this side of Christmas, I'm afraid. The best I can do is a direct flight on the 30th.' After more searching, he found a London to Zurich flight with Swiss Air that could connect with a Thai Airways flight to Bangkok, then another flight to Calcutta. That's the best I could do. I confirmed the details with Swiss Air, then went home to get some sleep.

"I sent a telegram to my betrothed telling her that I had missed my flight, but I would make it in time. The big day was fast approaching when we would exchange vows in front of an intimate gathering of 500 of our nearest and dearest friends and relatives. She said to her parents, 'See, I told you I shouldn't marry a foreigner!' She still reminds me of that.

"When I arrived at check-in next morning, I had an hour and a half to adjust the limit on my credit card, pay for my flights, check in, and board the aircraft to Zurich. When I did manage to connect with Mastercard, a very nice lady explained that though she appreciated my problem, she was only the office cleaner. Finally, I managed to speak to someone who could fix the card problem. I had just 10 minutes to catch my flight when the chap at the Swiss Air desk called Mastercard Authorizations.

"With a feeling of complete despair, I felt certain they wouldn't be able to approve it in time. But then, a miracle! At 08:20 the transaction was approved. Without saying a word, the agent printed my ticket, then leapt over the counter, clearing it in one practiced jump. Grabbing my bag, he sprinted toward the gate, beckoning me to follow.

"The flight attendants were clearing away the glasses in first class. The aircraft would leave on time. The door was about to close. As we ran to the gate, he yelled into his walkie talkie, 'Hold up, last minute passenger is desperate to make the flight.' At 08:29AM, my new best friend, the agent, sprinted down the jetway calling for them to hold the door. After that, it all seems a blur. My luggage was taken from me and placed into the hold. I was shown to my seat and the aircraft pushed back.

"My stopovers in Zurich and Bangkok were uneventful. I arrived in Calcutta with all my luggage and jet lag. I'm told I participated in the ceremony, but I don't remember much of it. I think I dozed often. My father would not speak to me for a week, though eventually I think he understood.

"On my return to Cambridge, I presented Hermann with the credit card bill. After reminding him that he had promised to get me to my wedding on time, he approved the claim for payment. Nothing more was said about it.

"Some of the Advanced R&D group moved back to 4A Market Hill after Acornsoft moved to Cherry Hinton. We set up a *skunk works* to build a new machine that sat outside everything Acorn had done before and was currently involved in. The Communicator combined many elements of the BBC Micro, but it used the WDC 65SC816 processor, which was the 16-bit version of the 6502. What we designed was a desktop computer dedicated to office communications applications, long before we'd heard about the Internet. It had an inbuilt telephone, modem and Econet interface for voice, data and file storage, as well as shared printing. Its support for extra memory and some multitasking abilities made this an interesting project, though with the ARM chip and Archimedes on the horizon, it seemed a little out of place.

"In late September 1985, I pulled an all-nighter because our first child was born. I had another sleepless night the following evening with the rest of our team, racing to finish the first Communicator, so it could be ready for an important sales presentation. It must have been a successful pitch because soon after that, Paul Bond set up a team at Cherry Hinton to write the operating system. Since the takeover, things had changed rapidly at Acorn – but from my perspective, not in a good way. It wasn't long before the members of our project team were retrenched, as part of Olivetti's 'business realignment'.

"Christopher moved a small group of us into the old servants' quarters at his home, Croxton Park, where we set up General Information Systems. I continued to work with him on smartcard development until late 1995, when I left to pursue other interests. I'm retired now but I still look back fondly on the excitement of this part of my career."

00011100

Acorn Downunder

Consolidated Marketing Corporation (CMC) was a small electronics business that imported and distributed anything that might turn a dollar. Bedhead clock radios were all the rage in the 1980s – particularly in hotels and motels. It was considered very *20th Century* to reach above the pillow and adjust the radio/alarm clock installed in the bedhead.

Handheld calculators were still something of a novelty. CMC was also Clive Sinclair's calculator distributor in Australia. Along came the ZX80 so CMC found itself in the computer business – supplied through Bob Bayham of Sinclair International. Christopher Curry was still with Sinclair Radionics at the time. He knew Bob well, so when Acorn wanted to find overseas markets, he turned to Bob for advice. Believing Acorn was no competitor in the antipodes, Clive gave Bob his blessing to set up – Acorn International. CMC began to sell Atoms and Eurocards.

I was serving as an avionics engineer at Air Force Headquarters in Melbourne when CMC's Managing Director, Julian Barson, first contacted me in 1981. Julian just learnt that his Econet wasn't the first local area network in Australian schools. I designed a network based on the Tandy TRS80 that had been selling since 1979.

It was an odd thing to admit, but Julian had doubts about the Atom's reliability and asked me to evaluate it for him. I spent the Easter break putting it through its paces. The conclusions were unflattering but honest. He read the report then phoned Hermann Hauser to introduce the two of us. Hermann agreed with my findings; then told us about the BBC Microcomputer, which was yet to be formally announced. It seemed to resolve the issues I had raised and offered a lot more features than anything else on the market at that time. Hermann asked Julian to post him a copy of my report. Apparently, I found issues that they missed.

By the end of the call, I decided to resign my commission and join Julian to set up Acorn Computers in Australia. He changed CMC's name to Barson Computers. It seemed like a great opportunity for both of us.

Unable to leave the service until November, I lived a double life for the rest of 1981: Air Force officer by day; setting up Acorn most evenings and weekends, starting with writing the tender bid to supply computers to schools in Western Australia. They heard so many positive comments about the BBC Micro from their UK contacts, the tender panel extended their deadline to allow more time to deliver a sample.

In May, I wrote a chapter about networking for *Microcomputers Don't Byte* – probably the first Econet material ever published. While there are technical differences between networks, they have a lot in common; so, most of it was drawn from working with my TRS80 network.

When John Coll sent me the galleys for an early draft of the *BBC Microcomputer System User Guide*, it was riddled with errors, so I returned a marked-up copy, with most pages covered with red-pen amendments. After sending off the package, there was a strong sense of editor's remorse. I had spoken to John a few times by phone but now worried he might resent the extent of my corrections. One evening, he phoned me; rather than being unhappy, he asked for help.

This was before email, so we spoke for hours on the phone, and used Telex to confirm our changes. The phone bill would have paid for a new Bentley, but it had to be done. A preliminary version of the User Guide came out (the blue cover version). We printed our own in Australia. Later, all copies were replaced with the official release from the UK.

By early November, I was on the payroll as head of the Acorn division of Barson Computers, which was little more than a startup at the time. Within an hour of rejoining civilian life, I took my first order for a network of BBC Micros with a File Server from Westbourne Grammar School. They wanted to get their order in first to be at the head of the queue.

Satisfied customers make the best salespeople. When schools called to ask about the BBC Microcomputer, we referred them to other schools already using them. Most called back within a few days asking for someone to come out and take an order. That's all it needed to make sales. By the end of 1982, despite slow deliveries from the UK, the order book showed well over 1,000 machines sold – all by word of mouth.

After settling into my new role in November, I flew to the UK to obtain a sample for the Western Australian Education Department tender. The first UK shipments were still weeks away, so it was going to be difficult, but Hermann and Christopher were both very supportive. Hermann told me to collect one of the first ten pre-production boards from their store to take as a sample. The bare boards had just arrived from Cleartone and Glyn Phillips was starting to populate them. Learning that my board wouldn't be ready before I returned home, I offered to help move things along, so we worked together to assemble them – a team effort.

Over dinner with Christopher Curry that last evening in Cambridge, we discussed what else was needed to assemble a working BBC Model B with disk drive and printer. The first item was the case. On cue, Allen Boothroyd arrived with the first successful sample from a factory near Manchester. Christopher examined it for a moment, then passed it to me. 'You need a case; well now you have it,' is all he said.

I also had a keyboard, but the power supply wouldn't be available until January. External supplies could be used for debugging. The workaround for that modification is still visible inside the case of #100002 where I drilled holes through the divider to feed the DC power cables.

At that time, John Cox was working on the disk filing system (DFS) with help from Paul Bond and Joe Dunn. There was nothing to do but wait. The deadline to present a fully documented, working sample was the end of March. It had to include the DFS, disk drive, printer and video – working perfectly. No pressure!

Back in Melbourne, I hooked up external power supplies and a green-screen monitor. It didn't work, but it was nothing that a day on the bench couldn't fix. By Christmas, the circuit board was working, but it lost video after a few minutes. The ULA struggled with our balmy December sunshine. A home-made heatsink fixed that. There were problems with the board. A trimmer capacitor and some jumper wires solved a couple of bugs. Each fault was reported by Telex to Acorn to assist with updates.

Acorn's focus was on its much larger UK market. Countries outside North America and Europe had to learn to be independent, inventive and flexible. When shipments arrived from the UK, most machines needed some repairs and adjustments. Making changes like converting the modulator from UHF to the VHF band wasn't a problem for us. If we hadn't been willing to modify UK machines, shipments would have been delayed for a year or more. For us, it was the only practical solution.

The BBC Micro's disk interface wasn't much different to the Eurocard System version, but it hadn't been tested on the board shipped to us. Shortly after Brian Cockburn joined the company, the first samples of the DFS arrived. We debugged it with long-distance help from John Cox.

Once the disk and Model B functions were all working, we shipped it to Western Australia for evaluation. It was obviously a prototype, but they were happy with what they saw. In June, the WA government announced Barson Computers as the exclusive supplier of disk-based computers for their schools. All other states and the Australian Capital Territory followed suit, reflecting the high regard in which the BBC Micro was held. Even Apple failed to achieve that distinction.

By late February, small shipments of BBC Model A machines began to arrive in Melbourne, so we could start to sort out the Econet interface. Joe Dunn and his team supported us brilliantly. Once Econet was operational (mid-1982) all schools that had been promised deliveries received their local area networks. Later that year, Tasmania and South Australia announced Barson Computers would be exclusive supplier of disk and network systems to their schools.

Literary Contributions

Getting user documentation from Cambridge was always a problem. Fortunately, the disk user manual for the BBC Micro could be adapted from the Eurocard System to become the DFS manual, but the Econet manual needed a lot more work. We had been shipping pre-release NFS EPROMs for a year, yet there was no sign of when it would be released in the UK. Prentice-Hall approached Joe Dunn to write a book about the Econet. I'd contributed a paper on network management, published in *You could use a Computer* in 1983, so he suggested they talk to me. I decided this could solve the need for operation and management manuals.

Joe appreciated the help we gave him, so he paid us a visit in August 1983. I flew back to the UK with him, which gave us time to do a deep dive into the evolving Econet specification. We sat up in the bubble of a Boeing 747 with nothing to do but enjoy airline hospitality and talk about networking. *Networking with the BBC Microcomputer* was published early in 1984. It was the first book to explain the Econet in detail and how filing systems 'talk' to applications using MOS calls.

Market positioning

Pricing wasn't much of an issue for Acorn because the UK's much larger population and smaller land mass meant their population density was one hundred times greater than Australia's. The high cost to reach the retail market meant it was never an option for Australia. Acorn had the BBC and the UK government behind it. One Pound Sterling was worth two Aussie dollars, and being a British company, Acorn was the darling of the media. All they had to do was sort out their production and supply issues, and they would move a million – which is just what they did!

For Barson Computers the first step toward establishing its brand in 1981 was to withdraw from selling the Atom and to replace existing machines with BBC Micros when they became available. While it hurt for a while, it would have cost a lot more in time and reputation if it had tried to keep Atoms in schools. This strategy was a boost to credibility.

Dropping the Atom also allowed time to reorganise the business so Acorn and Sinclair could run side-by-side in Australia and New Zealand.

Schools loved us, though there were inevitable comparisons with price differences between the UK and Australia. We did our best to explain what it cost to adapt the machines for disk and network operation, and the much higher overheads, whereas most machines in the UK were still running on cassette tapes or cheap floppy disk drives from Watford Electronics, and other third-party suppliers. Our customers wanted the best drives available at the time, and that's what Barson was selling.

Factory Rebuilds

Because the UK had excess stock of Model As and too few Model Bs, only Model As were delivered to Australia and New Zealand. They were upgraded to suit whatever the customer wanted: usually Model A plus 32 kilobytes of RAM and Econet or Model B with disk drives. File servers were standard Eurocard System 3s and 4s with an Econet interface.

For the first couple of years, every new machine had to be stripped down, modified, expanded, and debugged. Barson Computers was acting as an Original Equipment Manufacturer (OEM), not simply an importer. The crews in Melbourne and Auckland did a remarkable job considering what we were up against. ULAs came from the UK but common components were usually sourced locally. This was not to short-change Acorn; they had all the demand they could handle in their own market.

Acorn Education

From a product perspective, the Australian and New Zealand education markets bore little resemblance to the UK. We delivered disk drives almost a year ahead of the UK, and Econet two years before them.

The software our schools most wanted at the time was some sort of word processor. With Acornsoft View too far away from completion to be useful, Christopher gave me his blessing to modify the four-kilobyte text editor supplied with Eurocard Systems so it could run on the BBC Micro. I was never a big fan of the 6502, but I got it working and added a few features without much effort. We burnt so many ROMs our ganged EPROM programmer ran every day. It may not have been the greatest word processor, but I wrote *Networking with the BBC Microcomputer* using it. Reporters at the Mackay Mercury newspaper wrote their stories using that same text editor, formatted the text, then sent them off to a printer server connected to a typesetter.

The BBC Microcomputer's reputation for quality educational software was well justified. It shipped in both directions with Australasian software also being used in British schools. One outstanding software developer in Australia was a talented teacher at a school in Warrnambool, Victoria, the late John Hingston. John wrote a solar system simulator that was 25 years ahead of its time. It featured high-resolution graphics and text when most software was written for lower resolution screens.

Moving on

My role had become less relevant by 1984. I left Barson Computers to lead a team designing an electronic directory for our national phone carrier. It probably seems strange now but there was a time when we dialled a telephone operator if a printed phone book wasn't handy. After more than a year herding cats on that project, I was ready for a change.

For something completely different, I set up an advertising agency. This was just as the Apple Macintosh and desktop publishing came to town. Barson Computers, which had floated on the Melbourne Stock Exchange in 1984, was my first client. My first ad featured a BBC Micro and a teddy. The headline read: '*Two things your child will still value at twenty-one*'

Growing pains

Chuck Peddle, creator of the 6502, designed the Victor 9000/Sirius 1 – a business computer that was not IBM compatible. Despite that, the Sirius 1 was Europe's most popular 16-bit machine in the 1980s. Barson Computers became its Australasian distributor and built them from knockdown kits to meet local-content rules.

ACT, Victor's UK distributor, later released their own line of Sirius-compatible machines, the Apricot. ACT became one of Acorn's growing list of competitors in the UK when it released the F1e machine at a price similar to the Model B+.

Barson was appointed ACT's distributor in Australia and New Zealand. Apricot machines became the market leader for business computers. Later, as cheap clones entered the market, Barson-branded machines were sourced from Asia – fighting fire with fire.

During a visit from Chuch Peddle to an Australian computer show in that era, he commented when he saw these strange bedfellows on the Barson Computers stand: Acorn sitting side-by-side with Sinclair, Sirius 1, Apricot and cheap clones. He said, "I never get tired of seeing my creations on display." He was referring to having a BBC Microprocessor and an Acorn Electron – both running his 6502 microprocessor – alongside a Sirius 1.

For a time Barson's business grew rapidly in all market segments.

New Zealand

Once Julian Barson was confident that the Acorn business in Australia was under control, he started looking around for the next big thing. Setting up the New Zealand business occupied his attention for a while, operating in partnership but independently from Australia.

Doug Pauling had held a senior position with Motorola in Hong Kong. He and his old friend Julian set up Barson Computers New Zealand in September 1982. It was largely a mirror of the Australian business, with product ordered directly from the UK to save the added cost of transiting through Australia. This led to some duplication of effort with machines having to be modified to suit local requirements in both countries.

When Barson Australasia floated in 1984, the New Zealand business, which had been independent, came under the same umbrella. But what started as a strong and profitable business began to experience some serious setbacks.

By 1990, it was clear that Barson could not survive. After a failed attempt to sell the business to ACT, major creditors and bankers pressured the company to meet its liabilities. Acorn Computer Group plc's Managing Director, Sam Wauchope, threw Julian a lifeline: It acquired Barson Computers Australasia Limited after negotiating a debt write-down with Barson's bankers. All non-Acorn business ceased, but retrenched personnel were paid their entitlements.

After the takeover, Laurence Hardwick, spent seven years as Acorn's National Product Manager in Australia. Other UK personnel visited from time to time, though operations were largely left for Laurence and local personnel to manage.

In 1994, Acorn New Zealand moved into retail markets. Doug proved that one man's intractable problem is another's opportunity: Direct marketing was new to New Zealand. A sales promotion that generated leads in shopping centres sold more than 2,000 Electrons in six months.

Another retail campaign in 1994 targeted schools. Voucher promotions were also new to *the land of the long white cloud.* Acorn teamed up with New Zealand's largest supermarket chain. Over a three-month period, sales of NZ$240 million worth of groceries resulted in Acorn shipping more than NZ$3 million worth of hardware and software to schools.

Acorn Computers Australasia operated under limited UK control until just before Acorn Group plc was liquidated in 1999. In its last year of operation, Doug Pauling headed up all Australasian business, sharing time between Auckland and Melbourne offices – the last man to leave.

Fair exchange

The Australasian experience is a very different story to what happened in North America. Yes, that is a much larger market but launching into the entire country without first testing the water was an overreach. The decision to launch into the USA was not inherently flawed; it was the way it was done. A small footprint in a few states to begin with would have been far more prudent; and it probably would have been successful.

As for the fate of the first BBC Micro to come to Australia: Glyn Phillips gave #100000 to Christopher and #100001 to Hermann. After more than 12 years of faithful service, #100002 was turned out to pasture in 1994. Still bearing the scars of the quick-and-dirty modifications I made to it in the summer of 1981/1982, it is the oldest BBC Micro known to exist – now in the safe custody at the Museum of Victoria.

00011101

A Fairytale Ending

Superficially, it may seem that Acorn Computers changed little over its 20-year existence. In fact, Acorn was regularly engulfed by external forces over which it had little or no control, forcing it to reinvent itself each time. Hermann Hauser and Christopher Curry recruited clever people who helped them build the business. Acorn reached for the stars, not just once, but several times. Winning the contract to create the BBC Microcomputer and later the ARM RISC processor are two examples; but they were innovators and technical leaders in an era when others slavishly followed the market leaders.

Some observers link Acorn's highs and lows to changes of leadership, though the opposite is nearer the truth. At the top, at least, leaders changed in response to external influence. What remained constant for most of its history was the small group of very capable hardware and software engineers who stayed true to the ideals on which Acorn Computers was founded.

As for Acorn Group plc, its fairy godmother arrived in the form of a merchant banker, showed them how to have a happy ending. This is the last story...

ARM rising

Sophie is credited with naming the *Acorn RISC Machine* (the ARM chip). The company mustered all the resources it could to make it a success. Meanwhile, the Acorn Research Center in Palo Alto, California worked to create ARX – a world-class operating system. Perhaps it was hubris or perhaps the senior managers had forgotten their roots – Jim Mitchell and Carl Dellar led a talented team, but taking the project out of the Cambridge environment was a mistake. Sophie demonstrated this when she enlisted Paul Fellows to build the first successful ARM operating system – Arthur. That saved Acorn long enough to develop it into RISC OS. As the success of Arthur and the failure of ARX demonstrated: It's essential that business leaders never take the culture of the organisation for granted.

Many of us attribute Acorn's early successes to luck. It's good to be lucky but it still needs talent and hard work to make the most of opportunities. The ARM chip that Sophie, Steve and their team designed turned out to be more than they imagined, when its low power consumption created unexpected opportunities. Apple didn't want to use a product with Acorn's name on it, so a joint development and a name change from the *Acorn RISC Machine* to the *Advanced RISC Machine* led to the formation of Arm Limited.

It was another lucky break to have Apple on side. In November 1990, Apple invested cash, Acorn provided 12 engineers and VLSI Technology provided the tools. Advanced RISC Machines first operated in a converted barn at Swaffham Bulbeck, near Cambridge.

Robin Saxby, who had been Acorn's Motorola Semiconductor account manager in the late 1970s, was appointed CEO of the joint venture in 1991, He introduced a business model where the ARM processor could be licensed to any company for an upfront fee plus royalties paid on the number of chips produced. This effectively made Arm a partner to these companies, and it helped them reduce their time to market.

Ironically, Acorn won its second Queen's Award for Technology (this time for the ARM chip) in 1992 – two years after Arm as a separate business, managed to prise itself free from Acorn control.

The crucial break came with Texas Instruments in 1993: Nokia was advised to use an Arm-based TI processor for their new GSM mobile phone – the Nokia 6110. (You may remember spending hours playing snake on one.) Nokia was concerned about design issues that would increase the cost to produce. This led to Arm creating a custom instruction set that lowered the memory demands. This design was licensed by TI and sold to Nokia, where it was a huge success. The ARM7 became the flagship microprocessor for mobile phones, with more than 10 billion chips produced in 10 years.

This gave Arm Holdings credibility and proved the viability of the company's licensing model. In the traditions that had shaped Acorn, the relatively small size and dynamic culture of Arm gave it a response-time advantage in product development. All of its hard work came to fruition in 1994, during the mobile revolution when small mobile devices became a reality. The stars aligned and Arm was in the right place at the right time.

Acorn owned a large portion of the fledgling business and those who held shares in Acorn were hanging on to them to keep a stake in Arm. Stan Boland did a brilliant job of steering Acorn through the transition.

The Breakup of Acorn 1998-1999

While the relevant financial reports are available online through the Companies House portal, the moves and changes in company ownership, designed to minimise capital gains tax liabilities are complex. Read them at your peril. Here is a simplified breakdown of what followed.

In the first half of 1998, Arm Holdings plc completed a joint listing on the London Stock Exchange and NASDAQ with an IPO at £5.75. Acorn raised £18 million from the Acorn/Apple/VLSI Technology float. In June, Stan Boland replaced David Lee as CEO of Acorn Computers.

With losses of £9 million by September, the company underwent rapid restructuring to reduce ongoing losses. Announcing the changes, Stan outlined the direction Acorn would take: "The future of this company lies as a leading player in digital TV system components…"

The plan involved closing or selling the workstation division, reducing staff numbers by 40%, and cancelling the Phoebe project. It might have been saved but a draft memorandum of understanding concerning the revival of the Acorn PC was leaked and posted online. This may explain why Acorn's former technical director, Peter Bondar, didn't secure the assets of Acorn's product business for Applied Risc Technologies. Whatever the reason, Phoebe didn't survive the ordeal.

In October 1998, Castle Technology acquired the distribution rights for the existing designs of desktop machines to supply Acorn's dealer network. In January 1999, Apple Computer paid Acorn £3 million for its 50% interest in Xemplar Education, Finally, in March, RISCOS Ltd acquired a licence to develop and release RISC OS.

Acorn intended to focus on developing digital TV set-top boxes, high performance media-centric digital signal processors (silicon and software) and a reference design for a Windows NT thin client. The company hired a group of former STMicroelectronics silicon-design engineers and set up a £2 million silicon-design centre in Bristol.

By January 1999, Acorn Computers Limited had become Element 14 Limited (still owned by Acorn Group plc). The name referred to the fourteenth element in the periodic table – silicon. This change reflected the new business focus and distanced itself from the education market.

Pace Micro Technology acquired Acorn's set-top box division for about £200,000, along with Acorn's rights and service obligations.

In October 1999, NewJam Inc (founded six months earlier by Stan Boland and other senior Acorn management) purchased the silicon and software design business for £620,539, thus acquiring Element 14's name, staff and intellectual property. The following year Broadcom acquired Element 14 in an all-stock transaction for about US$600 million. Broadcom issued about 2.65 million shares, based on its price of US$224 at the time. This was done to fill a gap in its ADSL portfolio.

The ALARM team moved to NewJam Inc in 1999, later joining the Broadcom organisation. Sophie was appointed Director of IC Design in Broadcom's Cambridge, UK office, and was the Chief Architect of their Firepath processor.

Acquisition and asset disposal

Morgan Stanley proposed the means to release Acorn's 24.4% shareholding in Arm, while minimising potential tax liabilities to Acorn shareholders. A Morgan Stanley subsidiary, MSDW Investment Holdings Limited (a Cayman Islands bank) acquired Acorn Computer Group plc and its subsidiary companies. This released Acorn's £300 million shareholding "...using the purchase as a tax loss and swapping Acorn investors' shares for Arm shares..."

The steps followed sophisticated tax planning advice. In summary:

Acorn Group plc's stake in Arm was moved from its balance sheet into a subsidiary, Applied Risc Technologies, just before year-end.

MSDW, acting as purchaser, made an offer for Acorn Group plc.

The buyer structure allowed the Arm shareholding to be dealt with by employing share exchanges or other corporate reorganisation techniques rather than a direct disposal of the Arm shares for cash. The economic value of Arm was distributed to shareholders without any cash changing hands. Under UK tax law this becomes a tax-neutral reorganisation – avoiding a taxable capital gain.

MSDW retained about £47 million in Arm shares as a result.

The press reported the arrangement as an ARM takeover of Acorn even though ARM publicly disclaimed any involvement.

Morgan Stanley's London office is located at 25 Cabot Square, London E14 4QA. Not a lot of thought was applied to renaming Acorn Computer Group plc and Acorn Computers Limited.

In 1997, Acorn Computer Group plc was renamed Cabot 1 plc. It was delisted and renamed Cabot 1 Limited by MSDW Investment Holdings Limited in February 2000. Cabot 1 was dissolved on 9 December 2015.

After disposing of the Element 14 business in October 1999, the company, formed in December 1978 as Acorn Computers Ltd, was little more than a shell. It changed its name one last time to Cabot 2 Limited.

MSDW Investment Holdings Limited continued to administer the remaining assets of the business and to tidy up contractual and logistical obligations, including the servicing of product warranties.

In late 1999, Reflex Electronics signed a five-year contract to perform warranty work and support for Acorn-manufactured products, renewing an earlier arrangement with Acorn. Cabot 2 was dissolved on 16 June 2015.

The joint listing in London encouraged existing Acorn shareholders in the UK to have continued involvement. Arm stock soared, lifting this small British semiconductor design company into a billion-dollar business. This created the opportunity for Acorn shareholders to transfer their holdings to Arm through a two-for-five share swap. Since then, the value of Arm has continued to grow.

And if you are one of the fortunate few who held onto your Acorn shares throughout the magnificent highs and bravely weathered the storm during those desperately difficult lows: **May you live happily ever after!**

Further Reading

Bombs blast Israeli centres

SYDNEY: Two people were injured in a bomb blast at the Israeli Consulate in Sydney yesterday, and later two explosions rocked a crowded Jewish social club.

The Federal Government called in the Protective Services Co-ordination Centre in Canberra, and NSW police were alerted to guard synagogues and other Jewish centres.

Last night a man phoned the ABC newsroom in Sydney and said the Organisation for the Liberation of Lebanon of Foreigners was responsible for both bombings. He said attacks on Jewish properly would continue till Israel ceased to occupy Lebanon.

After the explosion at the Israeli Consulate in Westfield Towers, William Street, the Consul-General. Dr Moshe Liba. said police had informed him that a woman had phoned to say the Palestinian Liberation Organisation was responsible.

But a PLO spokesman in Melbourne, Mr Ali Kazak, denied his organisation had any connection with the explosion, which seriously damaged three floors of the building.

Westfield Towers accommodates parliamentarians and the offices of the Stewart Royal Commission into Drug Trafficking. It also houses the offices of two former Prime Ministers, Sir William MacMahon and Mr Whitlam.

Sir William was in his office on the 19th floor at the time of the explosion, and heard it. He said that about seven minutes later he was advised to leave the building by police, who said they were searching for another bomb. Mr Whitlam had been expected to arrive at his office soon after.

The bomb had been placed outside a door leading to the fire stairs, police said.

It went off soon after 2pm, blasting a 20-centimetre hole in the solid concrete walls, a 40-centimetre hole in the floor between the. sixth and seventh floors and blowing out all the fire doors up to the 10th floor.

A cleaner in her 60s who was working near a fire door when the bomb exploded was detained in hospital with head injuries, but her condition is believed not to be serious.

A man in his mid-30s who was working at his desk on the sixth floor was treated for shock at hospital before being discharged.

The explosions at the Hakoah Club in Hall Road, Bondi, badly damaged a number of cars in the underground car park, but no-one was injured.

They occurred just after 7pm, and about half an hour later the Cosmopolitan Motor Inn in Double Bay was evacuated when a phone call was received threatening another bomb explosion.

Late-night shoppers and club patrons were evacuated in Bondi and Double Bay, the eastern suburbs which are the hub of the Jewish community in Sydney.

In Canberra, police protection of the Israeli Embassy was increased last night, and the National Jewish Centre at Forrest was being guarded.

Canberra Times (ACT), Friday 24 December 1982, page 1

00011110

Special Mention

Basil Sands

Some people help to make our lives interesting. Take the unflappable Basil Sands from BBC Enterprises, for example…

On 23 December 1982, I received a phone call from Basil to discuss sales plans for the *BBC Computer Literacy Project* in Australia and New Zealand. As Head of BBC Enterprises in Australasia, he was responsible for marketing their TV programmes and merchandise. Basil hailed from New Zealand but behaved like the archetypal English gentleman: utterly unflappable, a more considerate but somewhat absent-minded man, you would be unlikely to meet.

A few minutes after taking the call, there was a loud muffled sound from Basil's end of the line. The sound was obviously an explosion. He seemed oblivious to the noise, dust and smoke, which newspaper reports later described as coming out of the ventilators; and the apparent agitation of his staff which he described as, "… a lot of screaming and yelling going on." and "Hold on while I close my door. I can't hear myself think." By now it was clear that we were in the Twilight Zone, but with a little prodding, he went to investigate. A few minutes passed, when my phone rang again. It was Basil.

It was one of the few terrorist attacks to take place in Australia. A bomb exploded less than 100 metres from his office on the floor above him; but it passed Basil Sands by with barely a moment's thought. His reaction was not indifference. He simply focused on what he could influence. I never knew him to waste time or attention on anything that was beyond his control. It was one of his more admirable qualities.

Note. The police failed to prosecute those responsible. The case was reopened in 2011 when Husayn Muhammed al-Umari (also known as Abu Ibrahim) was identified as the mastermind. Reports indicate he is still at large, wanted for this and other terrorist attacks.

BBC Microcomputer Resellers tunnel into bank

There is one more story – quite sad to relate. Barson Computers had little interest in selling through retail outlets, but two young men who had recently been released from prison, wanted to set up a BBC Micro retail store with encouragement from the Victorian Minister for Corrections. Julian Barson negotiated an arrangement where they could purchase computers at a large dealer discount in return for committing to minimum monthly deliveries.

They had paid for their past misdeeds and were trying to turn their lives around. One even completed a computer science degree while serving his sentence. After setting up their store and taking the first few months' deliveries, they found what we at Barsons had always believed: There wasn't a home computer market for the BBC Micro in Australia. Instead of asking for help from us or defaulting on the agreement, they reverted to old habits by trying to tunnel into the neighbouring bank – a cash withdrawal would allow them to pay their bills.

Caught red-handed, they spent more time behind bars. Such a waste. It would be easy to take a high and mighty stance in situations like this, but they really were two men who lacked the capacity to make sound life choices. Some things haven't changed much in 200 years. Australia was established by people who found themselves in similar circumstances.

Plan to dig in to bank, court told

MELBOURNE: Two men had almost tunnelled their way into a bank vault containing a large amount of money when they were arrested early yesterday morning after a week-long surveillance operation by police, the Melbourne Magistrates Court heard yesterday.

Detective Sergeant Ian Baker, of the Victoria Police major crime squad, said the two had dug the tunnel during the past month from a car park at the rear of the Commonwealth Bank in northern suburban Preston using a pick, a jackhammer and an electric drill tapped into a nearby street light.

Sergeant Baker said the tunnel had made an impact on the bank vault when the two were arrested at about 2am yesterday.

Before the court were Mr Joseph Marijancevic, 34, of Preston, and Mr Brian Royce Downs, 27, of Brooklyn, in Melbourne's western suburbs, both company directors of a computer firm.

Each was charged with conspiracy to commit burglary, trespassing with intent to steal, unlawfully possessing explosives and going equipped to steal on January 21.

Sergeant Baker said Mr Downs had admitted his part in the scheme, which had been undertaken to ease the liquidity problems of the pair's computer firm, in which he had invested $60,000.

Dctective Sergeant Kim West, of Preston CIB, said Mr Downs had been found down the concealed tunnel. Mr Marijancevic had been concealed in a laneway nearby with a two-way radio and had been arrested after a short, violent struggle.

Mr Marijancevic had later taken police to the pair's business premises in Preston, where a search revealed seven sticks of gelignite and 20 detonators which Mr Marijancevic had admitted were to be used to blow open any safes in the vault. Sergeant West said.

He said Mr Marijancevic had also said he knew there was a large' amount of money in the vault.

He said Mr Marijancevic had been seen in the car park the night before, when he had made an alteration to the tunnel entrance.

Mr Marijancevic told the court he totally denied all the charges and had never made the admissions police had said he had.

He told the magistrate, Mr Max Saunder, SM, that he had refused to take part in a signed record of interview with police.

No formal pleas were entered.

Canberra Times (ACT), Sunday 22 January 1984, page 3

John Coll

John A. Coll co-authored the specification for what would become BBC BASIC and wrote the *BBC Microcomputer System User Guide*. Sadly, John left us far too early and is sorely missed; everyone who knew him shared a genuine fondness for his enthusiasm, intelligence and vision.

He often spoke about *Le Centre mondial informatique et ressource humaine*. Sadly, as with other advanced ideas of that period, it is no longer with us. Drawing on its work as a backdrop to present his ideas, John spoke about a time when people would no longer rely on telephone books to look up phone numbers, and when air travel and accommodation would be arranged online, using a personal computer and a phone line. He described a time when we would have access to vast quantities of information at our fingertips – able to ask a question about almost any topic and receive instant, accurate answers. He also spoke about devices that would control homes, offices, vehicles, and more.

In short, John predicted: the coming of the Internet, eCommerce, Google, online booking systems, the Internet of Things, and more. It was his mission from the mid-70s to give flight to these ideas. Everything John predicted has come to pass. I wish he could be here to see just how right he was.

John visited Australia in 1983, then again in 1984 as keynote speaker of our national education conference. He arrived dressed head-to-toe in scuba diving gear – complete with tanks, mask and spear gun (not loaded). His keynote address covered many of the topics outlined above. He spoke about finding the courage to take risks, citing as an example, the chance that people might think him a complete idiot for dressing in neck-to-knee neoprene. If anyone did, they never showed it.

He was an exciting and interesting character who helped us see the future, long before any of his predictions had become mainstream ideas.

00011111

The Radcliffe Report

BBC Executive Producer, John Radcliffe, prepared a report in April 1981, two months after Acorn was selected. It provides his perspectives on the events that led to the *BBC Computer Literacy Project.*

BBC MICROCOMPUTER SYSTEM
THE CHOICE OF EQUIPMENT

This note sets out the background to the decision to choose Acorn Computers of Cambridge, as the providers of the BBC Microcomputer.

1. The Continuing Education Television Computer Literacy project, which has been in preparation since January 1980, is intended to provide a practical introduction for the layman to computers and computing. It was clearly necessary to provide for our audience to gain "hands-on" experience of computing for themselves; we had to think in terms of a "learning system", which included not only television programmes, books, and a linked course, but access to a microcomputer and appropriate software. Because of the current babel of incompatible language dialects, it would be necessary to base the project on a particular language specification; the only way to ensure that our audience would have access to a compatible machine at the right time was either to build the system around an existing commercial machine, or to license a BBC machine.
2. Simply to recommend an existing commercial machine to our viewers would have laid the Corporation open to charges of partiality which it would have been difficult to counter. It was therefore decided that the best course would be for BBC Enterprises to license a manufacturer to produce a BBC machine to our specification, to be marketed as the "BBC Microcomputer". This was bound to benefit the chosen manufacturer to some extent, but the Television Service took the view that provided that the company was not permitted to

link its name with that of the Corporation in any advertising, and provided that the machine had unique characteristics which made it more than simply a commercial product in BBC guise, this arrangement would be acceptable. It was agreed that the Corporation would publish the hardware and software specifications for the equipment so that other companies who wished to sell either hardware or software compatible with the BBC system could do so. The project has always been likely to stimulate public interest in computers and computing, and thus to benefit the British computer industry as a whole. This view has been shared from the outset by the Department of Industry.

3. In considering what type of machine we should be looking for, the C.E. project team took a good deal of outside advice, from bodies like Education Authorities, The Council for Educational Technology, the Department of Industry, the Department of Education and Science, and Micro Users in Secondary Education. This suggested that the range of interest in the project would be very wide, from the home-user who simply wanted to learn programming and compute as a hobby, up to the serious educational or business user. It was agreed that the machine needed to be low enough in price for it to be accessible to a large audience, but sufficiently powerful and expandable to permit a wide variety of use; ideally this should include access to new technological possibilities like telesoftware (the distribution of computer programs by teletext direct into computer memory). Because of the size and diversity of the computer industry there were a very wide range of companies which might have been willing to build a machine for us. Because of the need for 100% reliability within the limited time available before the launch of the project, and the need to confine consideration to a manageable number of alternatives, it was decided to go for a company already manufacturing or developing a microcomputer. Because the BBC is a public body, it was decided that this should be a British company, manufacturing in the U.K.

4. There were a number of microcomputers on the market at that time, but in the Spring of 1980, two new machines were launched, the Sinclair ZX80 and the Newbury NEWBRAIN, which looked considerably more attractive than other alternatives. The Sinclair machine was available on the market, at around £100; we were therefore able to use it, evaluate it, and consult various people who had used and tested it. It clearly represented something of a technological break- through, in its compactness, elegance of design, and cheapness. It did however have a number of intrinsic limitations, in particular In his early advertising Mr. Sinclair made large claims

for his machine, claiming that " ... you can use it to do quite literally anything, from playing chess to running a power station". This statement was, however, simply not true, and after various critical comments it was removed from the advertising. Our advice, at that time, from various users outside the BBC, both inside and outside the educational world, was that the ZX80 was excellent for the first-time user who wanted simply to learn the rudiments of programming a computer, but that because of its low power and lack of facilities, it was not capable of the range of uses we were looking for.

If the ZX80 appeared to be a breakthrough in terms of the cheapness and compactness, the Newbury NEWBRAIN was not only compact, but extremely versatile and potentially powerful. It had been designed from the outset with the business user in mind and was altogether a more up-market machine; it was however likely to be considerably more expensive than the ZX81. Since it was not yet available for purchase, we approached the manufacturers to ask for a machine for testing. They offered to produce a simplified and cheaper version of their machine for us, and it appeared that this could be produced at a price which would not be disproportionately greater than the ZX80. Their machine, however, was still in the process of development, and although we were able to see prototypes, we were unable to evaluate it properly. There were protracted discussions with the company, but (contrary to various speculative press articles) no commitment was made by the Corporation to Newbury.

5. By the autumn of 1980 a number of new and competitive low-cost microcomputers had come onto the market, and it was clear that our range of choicc was wider than it had appeared six months earlier. To resolve the issue, it was decided to make a formal approach to those British companies which on the best advice available to us seemed likely to be able to provide an acceptable machine. On December 23rd, 1980, a letter was sent by BBC Merchandising to seven companies, (Acorn, Nascom, Newbury Laboratories, Research Machines, Sinclair Research Ltd., Tangerine, and Transam) to establish whether they would be interested in providing the BBC machine (a copy of this letter is attached). Of these all but Research Machines expressed positive interest, and meetings were held with each to establish what they might be able to offer. A microcomputer from each company was tested by at least two advisers or members of the project team.

6. As part of this process of evaluation, a meeting was held with Mr. Sinclair, at Villiers House, on January 23rd. Mr. Sinclair demonstrated his new, as yet un-launched machine, the ZX81, which was rather more powerful and

advanced than the ZX80. There was a lengthy discussion, in which it emerged that Mr. Sinclair would be willing to provide a BBC version of this machine, with various improvements, but that he was unwilling to modify the ZX81 language specification and limited in his willingness to adapt the machine so that it could be interfaced with other equipment; he was sceptical about the possibilities for telesoftware. A minute of this meeting by Mr. Sinclair's secretary, which is substantially accurate, is attached. A subsequent meeting was held in Cambridge, on January 27th, at which Mr. Sinclair reiterated his reluctance to go very far in modifying his equipment to meet our needs; he stressed that he was the biggest single British manufacturer of microcomputers and that he was therefore setting standards for the industry. Mr. Sinclair's account of that meeting is given in the attached letter to David Allen (C.E. Project Leader) dated January 28th, 1981. At the same time, he wrote to Merchandising, making a formal bid and quoting a price.

7. On January 30th a meeting was held to assess the strengths and weaknesses of the contending bids. The assessing group included representatives from the C.E. Project Team, BBC Merchandising, BBC Engineering Designs Department, BBC Research Department, BBC Transmissions Department, and outside advisers from Micro Users in Secondary Education, and from the Department of Industry.

   Six machines were considered (Acorn, Nascom, Newbury, Tangerine, Transam, and Sinclair). The criteria used were as follows:

   - Hardware (power, facilities, expandability, ease of use)

   - Software (compatibility with existing languages, facilities, available applications programs)

   - The Company (manufacturing track-record, after-sales service record, financial base, willingness to modify its product)

   - Price

   After lengthy discussion it was decided that three of the six machines had clear advantages over the other three. Of the three eliminated, Transam and Nascom appeared to be uncompetitive both on technical grounds, and on price. It was also agreed that the Sinclair proposal was not as attractive an option as the three preferred machines, despite its very low price, on the following grounds:

   - Both the initial memory of the machine, and its capability for expansion were substantially less than available elsewhere.

- The software built into the machine did not have as great a range of facilities as that offered by other manufacturers. The ZX81 language specification had a number of features which limited its compatibility with other BASICS.

- The ZX81's capacity for interfacing with other equipment was more limited than offered elsewhere.

- To upgrade the Sinclair equipment to the level offered by other manufacturers would have required extensive development; moreover, Mr. Sinclair's willingness to modify his design to meet our needs was substantially less than that of other manufacturers.

It was agreed to eliminate Transam, Nascom, and Sinclair, and to have a further meeting to reconsider the proposals of the three other manufacturers. This was held on February 12th, and after lengthy discussion the Acorn equipment was chosen. (The Newbury design, though attractive, was unproven. The choice between Acorn and Tangerine was very narrow, but on a number of heads the Acorn was unanimously agreed to have the edge).

Following this meeting, letters were sent to the five unsuccessful contenders, notifying them of the BBC's decision, and offering to provide them with the final hardware and software specifications so that they would be in a position to provide compatible equipment. There have followed a series of detailed technical discussions with Acorn, during which they have agreed to make various modifications to the detail of both their hardware and software specifications, to meet our needs. (The software specification has now been made public; it was sent to the other manufacturers, including Sinclair, as soon as it was finalised). The resultant specification has attracted much comment, and it is widely agreed that the BBC machine will serve our audience well.

8. This view has been borne out by the choice of the BBC machine, after lengthy investigation and comparisons, as one of two machines for use by secondary schools, for which the Department of Industry are prepared to provide matching finance. (This was announced by the Prime Minister on April 6th).

John Radcliffe
Executive Producer,
Continuing Education, Television.
22.4.81

00100000

Before the Revolution

We live in a unique period of human history. From a time, long before Roman legions occupied Britain, until the beginning of the 19th century, there were few technological advances. In the past 200 years that all changed, though slowly at first. Even 150 years ago, Julius Caesar might have recognised aspects of life in Britain he had known; then the development curve took a sharp up-turn.

Every human living today, whatever their education or background, and wherever they live in the world, plays some part in the technological changes that began during the Second World War. While most of us have benefited from non-stop development, we've also contributed through our seemingly insatiable demand for *new* and *better*. In some of the most remote parts of the world, people may struggle to get potable water, but they still have ARM-based mobile phones and solar panels to charge them.

Reflecting on the IT revolution, there are many lessons to be learned – hopefully to avoid making the same mistakes in the future; and while the events of many decades ago may seem irrelevant today, we can all benefit from revisiting them.

The Acorn story began in the late 1970s but to understand how and why, it helps to step back to the early 1950s, when the foundations of the modern digital era were laid. Most of what we take for granted today: digital television, computers, mobile phones, global positioning systems, CAT scans, and so much more, would not exist had it not been for those first steps that coincided with the era of the baby boomers.

The coming of the mainframe

J. Presper Eckert and John W. Mauchly were, it could be said, the Wright Brothers of the digital computer. While some argue this point, they led development of the first successful general-purpose computer, ENIAC, during World War II, and were first to build and sell computers in the private sector, starting with UNIVAC I.

Like many technologists, they were inept business managers. They sold their technology to Remington Rand in 1950. Eckert remained with the company and its successors, but Mauchly resigned in 1958 and went into business as a consultant. Remington Rand produced UNIVAC I in 1951. Just forty-six of those machines were built. Soon, there would be others in much larger numbers, but from the perspective of the early 1950s, this was mass production. Pretty soon, UNIVAC was the name people thought of whenever computers were mentioned.

A UNIVAC I made history when CBS used one to predict the outcome of the 1952 US presidential election. Remington Rand staff wrote a program that based its prediction on results from key precincts. On election day, CBS News reporters phoned in the data to Philadelphia where data-entry clerks entered the results into the UNIVAC via three key-to-tape machines. The program compared all three versions to catch typing errors. At 8:30 PM Eastern time, based on 3.4 million votes, the program predicted an Eisenhower victory by a wide margin. Slow to trust the result, the CBS television announcers refused to call it, but later admitted they had been afraid to believe that a machine could forecast the future. How things have changed!

In the mid-fifties, a UNIVAC appeared as Cupid on Art Linkletter's syndicated *People are Funny* TV show. The machine was little more than a punched card sorter, but the idea that it could help young couples find true love appealed to viewers. We must admire the marketing people for finding a way to make computers seem sexy.

Along came Big Blue

The ever-increasing need to capture and process data has its roots in the late 19th century. Well before the advent of electronic computers, Herman Hollerith developed electromechanical tabulating machines. Following decades of mergers, acquisitions, and name changes, Hollerith's business became International Business Machines (IBM, also known as *Big Blue*) in 1924. Its punched card and punched paper tape machines preceded its move into digital computers. They joined the race to build faster, smaller, cheaper and more reliable computers, with their first offering – the IBM 701 – shipping from 1952.

IBM has been the world's largest and most influential computer company, both in sales and in product performance. Its impact on Acorn includes Econet, which relies on IBM's SDLC protocol, and an IBM 801 research paper, which helped to kick-start Acorn's RISC development.

This summary of the IBM 801 helps to explain RISC concepts.

IBM 801

Registers and accumulators are the engine room of a computer's central processor unit (CPU). To achieve faster and easier-to-program computers, a rich set of register-to-register, memory-to-register, and memory-to-memory operations became common in the design of CPUs. This approach led to the development of complex instruction set computers (CISC). In a CISC machine, a single instruction like ADD might have a dozen variants, such as: Add two numbers in internal registers, add a register to a value in memory, add two values from memory, etc. This allows the programmer to select the exact operation needed for a particular step in the program.

On high-end machines, these complex instructions are usually implemented directly in hardware, while on smaller, low-cost machines they are simulated using a sequence of special instructions called microcode. This extended the simple fetch-and-execute cycles of the early digital computers. Tracy Kidder's Pulitzer Prize-winning novel *The Soul of a New Machine* does a great job describing its development for the Data General MV8000 Eclipse minicomputer. Microcoding is also used in microprocessors, such as the Intel 8086 and 8088 and Motorola 68000.

Microcode allows a processor to perform complex instructions without bloating the CPU with dedicated circuitry. The processor uses microcode to break it into a series of steps. The idea of offering all possible addressing modes for all instructions became a common goal for processor designers – the 'orthogonal instruction set'.

The efficiency and value of microcoding wasn't questioned until the 1970s, when an IBM research team led by John Cocke collected data on the performance of real-world workloads on their machines. They found that half of the instructions performed in typical programs are of five simple types: load value from memory, store value to memory, add fixed-point numbers, compare fixed-point numbers, and branch based on the result. They found that a simpler processor design is more efficient than CISC architectures. This defied accepted practices of the day, which were (and still are) based on using microcode – a technique that has long been championed by IBM.

On high-end machines, many of these instructions are implemented directly in hardware, like a floating-point arithmetic unit. On low-end machines some functions may be simulated using a sequence of special instructions in microcode.

John Cocke's team built the IBM 801, a machine that dramatically simplified CPU design. They achieved this by eliminating instructions that operated on data in main memory, leaving only those that worked on the internal processor registers. They also omitted having a floating-point unit. The result of this work was a reduced instruction set computer (RISC) with greater performance than its CISC cousins. This increased the number of instructions that could be performed per second.

The 801-project team noticed a side-effect: When faced with different versions of an instruction, compiler authors would usually pick the simplest version to suit the low-end machines. This ensured that the machine code generated by the compiler would run as fast as possible on the entire series of machines (e.g. IBM System/370). While other versions of instructions might run faster on a machine that implemented them in hardware, compiler authors chose to achieve cross-platform compatibility; so, many instructions available in CISC machines were never used in compiled programs. It was here that the 801-project team found that microcode imposes an expensive overhead in performing the most frequently executed instructions.

Using microcode takes extra time to examine the instruction before it is performed. Removing it eliminates this overhead, so compiled programs run faster. As microcode runs small subroutines dedicated to a particular hardware implementation, it was performing the same task as the compiler, while implementing higher-level instructions as a sequence of machine-specific instructions. Removing the microcode and implementing the steps in the compiler can deliver a faster machine.

One concern was that programs written for a RISC machine would take up more memory. Tasks performed with a single instruction on a CISC machine would require multiple instructions on RISC. For instance, adding two numbers from memory would require two load-to-register instructions, a register-to-register add, and then a store-to-memory. This could potentially slow the system if it had to spend more time reading instructions from memory than it formerly took to decode them.

One way to save time is to use a pipeline in the CPU. This pre-loads commands and/or data during the execution of one instruction, ready to execute the next. Also, as compilers improved, overall program length fell, eventually becoming roughly the same as those written for the CISC machine. As the ARM and other RISC architectures have consistently demonstrated, there is much to be gained from this design approach.

The IBM 801 is considered the first RISC-based computer. John Cocke's work was recognised with several awards, including the Turing Award, National Medal of Science and the National Medal of Technology.

This machine was one of the influences that led to Acorn's development of the first ARM processor.

International Computers Limited

International Computers and Tabulators (ICT, later ICL) was the UK's answer to IBM. Its predecessor had exclusive rights to sell IBM machines within the *British Empire*, but IBM's terms of trade were too onerous, so they released IBM from its watertight agreement. Had they kept it in place, it would have locked IBM out of the Commonwealth, but that's unsporting, so with a few kind words, they went their separate ways.

Author's Note. Computing began for me in 1967, operating an ICT 1905A mainframe. The sight of magnetic tape decks dancing back and forth was mesmerising; but even more impressive was the line printer. A drum containing its alphanumeric character set rotated 20 times per second, activating up to 160 hammers per line in sync with the drum's position, then advancing the paper to the next line. Modern laser printers outperform drum printers, but I miss the busy sound of that machine. In full flight, it delivered a constant drumroll. As slaves to the machine, we loaded boxes of paper into the feeder, then collected and dispatched the printed output. For me, 'batch jobs' were usually motor vehicle registration renewals, or designs for new highways.

ICL had struggled to change from building large machines to home and small business computers. Looking more like a sheltered workshop than a state-of-the-art facility, the BBC Microcomputer production line at the ICL plant in Kidsgrove wasn't set up for low-cost, high-volume fabrication. ICL had some modern production plants, but this empty shell wasn't one of them. It was clear why Japan and Hong Kong had overtaken the UK's manufacturing capability so quickly. Taiwan, China, Malaysia, Thailand and Viet Nam would soon follow.

While the BBC Microcomputer was based on the same principles as mainframe computers, they were entirely different beasts. ICL had agreed to build the BBC Model B, though this would be a short-lived activity as its heart just wasn't in it. After a less-than-stellar trial run, where they made about 30,000 units, Acorn and ICL parted company. Even pressure from the Department of Trade and Industry wouldn't sway them. Cleartone, AB Electronic Systems and Race Electronics in Wales, Keltek in Kelso and Wong's in Hong Kong built the next million.

ICL had built world class computers. They were in the business of 'serious computing'; but that was about to change. Events overtook ICL's fortunes, and it was acquired by Fujitsu in 2002.

How semiconductors changed everything

In the late 1950s, capital was readily available in the USA to anyone who could demonstrate their ability to make semiconductors. The promise of rich rewards led to a rush of investors willing to back new ventures. Access to capital enabled the rapid expansion of this fledgling industry; but having the knowhow did not guarantee success, which is why so many firms went bust – some spectacularly.

Many who led the charge designing semiconductors were very small operations. A few were little more than cottage industries in the early 1960s. The chain-of-command from the engineers on the shop floor to the CEO was short. When production was measured in hundreds and thousands, the financial impact of a bad decision was not as devastating, so decisions were made quickly. How things have changed!

The US semiconductor industry began in Massachusetts when three researchers at Bell Labs – Shockley, Bardeen and Brattain – invented the bipolar transistor in the 1950s. Shockley returned home to California. He set up Shockley Semiconductor Laboratories at 391 San Antonio Road, Mountain View. The original building has been demolished but this location is credited with kick-starting semiconductor production in what became known as Silicon Valley.

Finding Shockley impossible to work with, his brightest minds (the *Traitorous Eight*) left to set up Fairchild Semiconductor in 1957. The companies that would later be set up by Fairchild's ex-employees are known as the *Fairchildren*. Intel and AMD are two examples.

Fairchild and Intel were the first to combine multiple bipolar transistors on a single silicon wafer. This led to small scale integration (SSI) with fewer than 10 transistors, and medium scale integration (MSI) with 10–100 transistors, in the 1960s. Bipolar transistors rely on electrical current flow. While the current is small in each transistor, cram enough of them into a slice of silicon (the substrate) and heat dissipation can become a problem, limiting the number of gates that can be packed onto a single chip. Field effect transistors (FET) rely on electric fields rather than electric current for switching, so the heating problem is much less. Julius Lilienfeld patented the FET in 1925, but it took almost 50 years to become commercially viable. Large scale integration (LSI) had to wait until a method could be found to fabricate these low power metal oxide semiconductor field effect transistors (MOSFET). By the 1960s, LSI devices with up to 100,000 transistors were being produced.

By the early 1970s, the goal was to cram more circuitry onto a single chip. It was a scramble, with several companies trying to claim firsts. Intel was just three years old when it announced the 4004 processor. The chip was only four bits wide. It was only used as a calculator chip, but it signaled the rewards yet to come. Clive Sinclair was one of the first to capitalize on this advance in chip fabrication with the Sinclair Executive calculator, using a chip from Texas Instruments.

Microprocessors miniaturize the central processing unit (CPU) of the mainframe. Add a crystal oscillator, input/output interface, memory with buffering and decoding: You have a microcomputer. Most of that can now be packed into one device – a system on a chip (SOC).

It is widely accepted that the first four microprocessors available in commercial quantities were the Intel 4004 (1971), Intel 8008 (1972), Rockwell PPS-4 (1972) and the first bit-slice processor – the National Semiconductor IMP-16 (1973). After these, the order of appearance is less clear as so many new microprocessors were released in quick succession. With the level of activity in the USA – particularly in Silicon Valley and Texas – the often-quoted Moore's Law was becoming a certainty. Gordon Moore observed that the number of transistors able to be fabricated on a chip was doubling every eighteen months to two years.

By the mid-1970s Intel, Fairchild Semiconductor, Texas Instruments, National Semiconductor, MOS Technology, Motorola, Signetics, and others were shipping very large-scale integrated circuits (VLSI) that contained millions, and later billions, of transistors, implementing complex digital and analogue systems.. Some semiconductor companies rose to dizzying heights and continue to be leaders today, while some disappeared almost as quickly as they arrived. Things were moving fast. The National Semiconductor SC/MP (Simple Cost-effective Micro Processor) was released in April 1976. Science of Cambridge used this chip to develop the MK14, its first microcomputer, a few months later.

From 1978, Acorn designed and built 8-bit computers. Within five years, two of its brightest stars, Sophie Wilson and Steve Furber, led a small team who created a new type of microprocessor (the Acorn RISC Machine, which later became known as the Advanced RISC Machine – the ARM chip) that would establish them both in the history books, and would create a multi-billion-pound business. Acorn's decision to design a RISC processor coincided with development initiatives across all areas of computers and communications. Rapid advances in semiconductor technology made it possible.

Post-war research and development in electronics, computers and communications in the USA changed the way people lived – pushed along by a strong economy and a can-do belief in themselves as a nation. Meanwhile, Britain was still recovering from the sacrifices they made to withstand the assault of Nazi Germany. The business environment was slow to recover, stifling innovation, and early advances in computing were hamstrung by a lack of capital and trained computer engineers.

In time, economic confidence improved in mainland Europe and the UK. Some companies produced low- and high-power discrete transistors, as well as memory and custom semiconductors. A few produced wafers, and some made complex systems used to fabricate integrated circuits. European companies seldom went head-to-head with US competitors, unlike Asian businesses that beat the Americans at their own game.

Today, Taiwan is the market leader in all areas of chip fabrication, though concerns about political stability in East Asia are changing the balance. Germany, the Netherlands, India, South Korea, Vietnam, Malaysia and the Philippines are all vying for a larger market share.

Uncommitted Logic Arrays

Digital circuits used a lot of SSI and MSI components. Real estate on a printed circuit board is expensive. Designers looked for ways to reduce the chip count. Uncommitted logic arrays (ULA) appeared in portable CB radios and cameras from the 1970s. As new fabrication methods improved chip size and performance, they began to be used in computers in the early 1980s. A type of gate array, they became big news in 1981 with the IBM 3081 mainframe whose design was based largely on them.

ULAs allowed Acorn to compress many integrated circuits into a few custom chips in the BBC Microcomputer, the Tube interface for second processors, and the Acorn Electron. Manufacturing gate arrays reduced production times and one-off engineering costs as fewer masks were needed. Unfortunately for Acorn, they learnt too late that they were pushing the technology beyond what was prudent.

Ferranti made ULAs for Acorn and also worked with Clive Sinclair around that time. An enhanced Sinclair ZX80 using a ULA became the Sinclair ZX81; another was used in the Sinclair Spectrum. Commodore, and other manufacturers of home computers followed suit throughout the 1980s – the Commodore Amiga, for example.

When gate arrays were chosen for the IBM PC, production numbers made semi-custom chips more economical, but their use declined by the end of the 1990s, replaced by other devices such as field programmable gate arrays (FPGA), complex programmable logic devices (CPLD), and other application-specific integrated circuits (ASIC).

Elements of a ULA

ULAs were prefabricated by laying down large numbers of transistors onto a semiconductor substrate. They are said to be 'uncommitted' at first, as their transistors aren't electrically connected to one another. The custom circuit is converted into overlays that represent connections between the transistors and buried 'tunnels' to make NOR logic gates. Applying De Morgan's Theorem, combinations of NOR gates can represent fundamental logic devices such as inverters (NOT), and OR, AND, NAND and XOR (exclusive OR) gates, as well as more complex devices such as flip-flops and registers.

Linking was performed by adding metal interconnects to unpopulated ULA substrates in Ferranti's fabrication plant. Masks specified how the connections were added to make custom circuits.

Two ULAs were chosen to meet the BBC Micro's size and cost constraints: one for the video display and one for the serial I/O (RS-423 and cassette interfaces). A third ULA was produced for the Tube interface – fitted to all second processors. Later, the Electron's ULA replaced dozens of the BBC Micro's discrete TTL chips into one custom chip. By unifying memory management, video, sound, I/O and system timing, it drove component count and cost down.

How the chip was designed

Steve Furber, Sophie Wilson and their team of engineers designed and tested prototypes using discrete components for the circuits needed to be laid down in the BBC Micro's ULAs. The job of laying out the chips was passed to Jeremy Dion and Peter Robinson at the University Computer Laboratory who had some experience with semi-custom chip design. They marked their contribution to the BBC Micro, by placing their initials (PR+JD) in metal, 50 microns high on the chips.

The metal tracks were about five microns wide, so an oscilloscope probe could be placed onto them for debugging. Andy Hopper borrowed a microscope and micro-manipulator from the Zoology department next door (which used them for probing insects' brains). To provide a stable optical bench, he bought a concrete paving slab from a local garden centre. The tip of the manipulator probe was chemically milled to a fine point. It carried a large capacitive load which made reading timings tricky, but not impossible.

The Serial Processor, Video Processor and the Tube ULAs contained hundreds of cells. Design software for those chips wasn't available in the early 1980s, so the masks were produced manually. The artwork was digitised by Ferranti, then checked by Acorn and the Computer Lab. Ferranti later produced a larger ULA that was used for the Electron. Reflecting the rapid advances in Acorn's understanding of chip design, Sophie Wilson developed artwork layout software on a BBC Micro. They would soon be designing their own RISC processor.

Ferranti bipolar ULAs drew more current than FETs. A typical Ferranti 440-cell design used about half its available components. At that density, the heat generated was manageable. Acorn's video ULA used more than 95% of its components; so, it ran a lot hotter than any other design that Ferranti had fabricated. Thus, the first samples could only operate with a heatsink.

Each cell contained a current source and ground connection, two pairs of bipolar transistors, two resistors and three 'cross-unders'. A single layer of metal was used to connect the components into NOR gates (two inverters or two-input gates, or just one three- or four-input gate per cell) and the gates connecting them to the desired circuit.

The biggest drawback to bipolar logic is that it converts relatively large amounts of electrical energy to heat, compared to MOSFETs. In the case of the BBC Micro, this led to heat dissipation and other problems that delayed early deliveries.

The prototype chips worked for about 20 minutes before overheating. That seemed promising enough for Hermann Hauser to order a batch of 5,000 chips and metal heat sinks to keep them cool. Unfortunately, the yield was less than 10% as Ferranti's chemistry was inconsistent, whereas the prototypes had been made on an unusually fast batch of silicon. They made a second batch with adjusted chemistry, and all was well, apart from the over-heating.

Acorn used a Ferranti 990-cell array for their second processor Tube connections. Despite heating problems, the Tube ULA worked well for several second processors, including the first-generation ARM1 chips.

Bipolar ULA fabrication has largely been replaced by CMOS gate arrays. Today, FPGAs, CPLDs, ASICs and other customisable logic devices have mostly replaced ULAs as engineers can create solutions without needing a factory to etch the design onto the wafer. That's progress!

00100001

BBC BASIC – A Primer

The British public was introduced to BBC BASIC in February 1982, when *The Computer Programme* was first broadcast on BBC Television. Timing is significant as this was just a year after the evaluation team met with Acorn engineers for the first time to look over a hastily built prototype. It serves as a great example of what can be achieved when people work together in a spirit of cooperation.

The BBC project team needed the right programming language for its computer illiterate viewing audience. BASIC – Beginners All-purpose Symbolic Instruction Code – had been created at Dartmouth College in the USA in the 1960s to introduce computer programming. Writing a BASIC interpreter was relatively easy, but writing one where there were no widely accepted standards was a different matter. Without standards and a governing body to control its development, it had been copied and changed so many times, there was little cross-platform compatibility.

The *Computer Literacy Project* team started with clear ideas about what its specification should contain. It had been shaped by the MUSE recommendations for ABC BASIC, as promoted by John Coll.

The BBC wanted to have an interpreter that could easily run programs written in Microsoft BASIC. This would allow viewers to enter program listings from computer magazines, which is how a lot of self-taught programmers started. But even Microsoft BASIC had so many different versions; and as popular as it had become, some important features were missing – including long variable names and structured programming. It would be some time before Microsoft added those features.

As well as compatibility with the market leader, structured programming was a top priority. Stepping away from the main flow of a program using an unconditional branch statement (GOTO), encourages programmers to write 'spaghetti code' as flow can go anywhere, rather than following a step-by-step sequence. *Purists* don't allow GOTO – ever. *Pragmatists* discourage the use of GOTO but are less set in their ways.

When Professor David Wheeler at the University of Cambridge invented the subroutine branch in the early 1950s, that helped. A simple GOSUB would take the flow to a different block of code and when completed, RETURN to the instruction after the point where it had branched.

Next came ALGOL in the late 1950s. It put structured programming on the map as code flowed in a straight line while enabling branching and looping. Most languages now support 'constructs' like: If-Then-Else, While-Endwhile, Do-Until; and there's BASIC's For-Next-Step. The syntax varies slightly among languages, but most support them now. So, learning programming with a structured BASIC is a great place to start.

Acorn had a well-developed language specification by late 1980 that had been called *Super Basic* by the BBC. When compared to ABC BASIC there were some notable differences. What followed were negotiations to finalise the specifications of BBC BASIC. Compromises were made on both sides. It was agreed that most programs written for one of the more recent versions of Microsoft BASIC should be able to run on BBC BASIC, while programs written specifically for BBC BASIC could take advantage of its more sophisticated features.

The BBC's decision to adopt their own specifications for BASIC proved challenging to Sophie Wilson (Acorn), John Coll (MUSE) and Richard Russell (BBC). Sophie was the creative force behind Atom BASIC and BBC BASIC. She also went on to develop BASIC for the ARM processor as used in the Archimedes series of machines.

There were differences of opinion about how to design a structured BASIC language, but what she produced has a strong following to this day. Some of its most popular features are its support for structured programming principles, speed, ease of use, direct links to the operating system, and its ability to embed assembly language programs (a brilliant way to learn and write machine code), all within BBC BASIC programs. The level of sophistication that can be achieved is extraordinary. For example:

Steve Furber wrote the ARM's first behavioural model, with its instruction set, microarchitecture and operation defined in 808 lines of BBC BASIC code. Engineers wrote test programs to check the behaviour of each ARM machine code instruction, and they wrote some ARM instruction simulators.

Derek McAuley wrote a hard disk printer spooler in BBC BASIC that scheduled print jobs for Acorn's large laser printer.

The Electron's uncommitted logic array (ULA) was to be the largest Ferranti had ever attempted, needing around 2,000 gates. As Jon Thackray has explained, Acorn had no simulation or synthesis tools, no clock domain crossing (CDC) checkers, or anything else that might have made the task seem more realistic, just Sophie's layout tool that was a large BBC BASIC program.

With its incredibly complex specifications, being able to anticipate how the 'as-built' ARM CPU would perform was a colossal challenge. An emulator written in BBC BASIC enabled the designers to test the sample chip when it arrived. The software verified that each component worked correctly.

Imagine loading untested software onto untested hardware, Jes Wills asked: 'What would we do if none of the lights came on?'

BBC BASIC provided the solution to all of these and countless other problems.

There was a time when learning to write computer programs was within the ability of most people. The code might not always be pretty, but the basic ideas could be learnt in a few hours. Simple steps performed one after the other in sequence are as easy to read as Aunt Bertha's recipe for sponge cake. But unlike her instructions, which should never vary, there are times when computer programs need to loop or branch in response to test conditions.

Today, most professional programmers would argue that the correct way to design code is using object-oriented programming (OOP), which incorporates structured programming within its own complex rules. To become proficient, the first barrier to entry is the seemingly endless stream of OOP ideas: Abstraction, Encapsulation, Inheritance, Classes, Objects, Overloading and Overriding, Polymorphism, Wrappers, Constructors, and more…

Object-oriented programming dominates professional software today. Java, C++, Python, and many other programming languages are built on an OOP foundation, but it's challenging for first-time programmers.

Both approaches have their strengths. A structured BASIC program could be written using OOP principles but it's less accessible to entry-level programmers. They may be wise to delay taking an OOP approach until they are ready to write more complex and advanced applications.

…

Chris Barrett's experience with BASIC

Chris is a volunteer at the National Museum of Computing. Like so many of his generation, the inspiration for Chris's career path as a professional programmer was learning to write programs. Here is Chris's story…

"I started programming in Sinclair BASIC at home, and later in BBC BASIC at school, before my family purchased an Acorn Archimedes in my teens. While Sinclair BASIC was arguably the most common programming language used in UK homes during the 1980s, a young me found it frustrating to work with, largely because of its one-keyword-per-key method of entering programs. I doubt I would have persevered with programming if Sinclair machines had been my only option.

"Most kids started programming by typing program listings from computer magazines, then spending hours figuring out why they wouldn't run; and most machines had some form of Microsoft BASIC when I was a kid, but it didn't seem to occur to them to set some standards. Even different models of the same brand of computer often had language differences.

"A friend of mine owned several Commodore machines, whose versions of Microsoft BASIC were rarely compatible with the listings I found in magazines and programming books from my local library.

"As well as the compatibility problems, the Sinclair and Commodore machines were painfully slow. It was almost impossible to write games that ran smoothly without resorting to assembly language, which required additional software tools, or hand-assembled machine code that had to be entered using PEEK and POKE commands. The learning curve was steep, and they were seldom compatible across platforms. One thing kids liked to do was share their latest game with friends. Unless the machines were the exact same model, that was out of the question.

"By contrast, BBC BASIC was fast and elegant. Keywords could be typed directly, rather than selecting them from a crowded keyboard. Pascal-like structured programming constructs such as DEF FN, DEF PROC, REPEAT…UNTIL and WHILE…ENDWHILE encouraged me to develop good programming habits from the start. These reduced the need for GOTO statements which made my code easier to understand, and which I now understand contributed to improved performance.

"The inline assembler was a great feature, allowing pre-teen me to write simple machine code programs using mnemonics instead of having to poke hand-assembled bytes (which were more difficult to understand) into memory.

"The BBC Micro also had a well-documented operating system that was especially easy to use when it had a disk-based filesystem. My Spectrum and my friend's Commodores were tape-based so even saving and loading programs was tedious and time-consuming.

"BBC BASIC felt professional, reliable, structured and thoughtfully designed, whereas the Spectrum felt more like a toy. My programming education was helped along when BBC BASIC was included on the ARM-based Acorn machines. I wrote small WIMP applications and assembled simple ARM code directly inside BASIC programs, with very little effort. To me, BBC BASIC was a fantastic learning tool and excellent general purpose programming language.

"I think BBC BASIC is as relevant now as it ever was. Just attend a BBC Micro retro computer meeting to see why people still love it."

00100010

Cambridge v Manchester

There have been times when Great Britain could rightly claim to be an intellectual juggernaut. Michael Faraday gave us electrical power generation and electric motors to drive the wheels of industry; then James Clerk Maxwell developed his famous equations to explain the how and the why of Faraday's work. Later, J. J. Thomson discovered the electron. Britain adopted New Zealander, Ernest Rutherford, as one of its own. After a stellar career in nuclear physics, he was buried near the graves of Charles Darwin and Isaac Newton at Westminster Abbey. Britain's dominance in the physical sciences led to its influence during the industrial revolution when James Watt was supported by legions of physicists, chemists, materials engineers and mathematicians.

Cambridge was home to many of the intellectual superstars of the 1800s and 1900s. The Cavendish laboratory played host to Maxwell, Thomson, Rutherford, and others. Rutherford actually split the atom while at Manchester, leaving some radioactivity around to be cleared up. But then he moved to Cambridge and made bits of that radioactive too!

The Lucasian Chair of Mathematics has been occupied by men such as Isaac Newton, Paul Dirac, Stephen Hawking and Charles Babbage.

The past 150 years has experienced an explosion of technology. The first 75 years were dominated by physicists, chemists, metallurgists, and civil, mechanical, electrical and radio engineers. The second 75 years has been the era of computer scientists, hardware and software engineers, programmers, analysts and a host of other new careers and disciplines that began as the baby boomers were born in the early 1950s. They were the generation who kicked off the digital revolution.

Babbage is regarded as the first computer engineer (with Countess Ada Lovelace being our first computer programmer). His inventions were a failure at the time and a drain on the coffers of public and private investors, but that didn't reduce his status as a pioneer. The history of Acorn Computers and its associated companies demonstrates that success and failure can take many forms and that neither must be terminal. If Charles Babbage was here today, I'm sure he'd agree.

Among British universities, Oxford, Cambridge, London, Manchester and Edinburgh are the best known. In the field of computer architecture, Manchester and Cambridge have dominated and have been centres for Britain's computer industry. Spurred along by rapid development during World War II, both centres established themselves as pioneers.

Their lists of outstanding contributors continue to grow, but the names from the history books who are most often recognised are: George Boole, Alan Turing, Maurice Wilkes, Roger Needham and Karen Spärck-Jones, David Wheeler, Freddie Williams, Tom Kilburn and Max Newman.

Who is better qualified to discuss the history of computing in the United Kingdom than Professor Emeritus Stephen Furber?

UK Centres of Computing History

"Born and raised in Manchester, I spent 20 years in Cambridge, at the university and at Acorn before my appointment to the ICL Chair of Computing at the University of Manchester. It should come as no surprise that I have shared allegiances to both cities.

"I wasn't aware of any tension between the two universities before I joined Manchester in 1990. Sure, there was a sense of competitiveness, but I didn't appreciate the level of ill-feeling, dating back to the early days of computing when both vied for the same very limited public and private funding. On the surface, it's expressed through debates about who really built the world's first stored-program computer. As with a lot of questions like this, there is not a simple answer, but in my view...

"Manchester built the world's first operational stored-program computer, but it was just a prototype and not very useful. Williams and Kilburn were not focused on building a computer, they were trying to solve the problem of building a larger memory storage device using cathode ray tubes, so they built a simple computer around their storage device to test the concept – the Manchester Baby in 1948. A year later, they built the Manchester Mark I. It was an enhanced Manchester Baby, that added index registers and a hardware multiplier. Five more machines would follow including the Atlas in 1962.

"The world's first stored-program computer that could deliver practical computing services was EDSAC, built by Maurice Wilkes at Cambridge in 1949. It was based on an architecture he designed onboard ship while returning from the USA, after attending a conference where the stored program computer concept was unveiled. Maurice hadn't copied a US design as none had been built, though the concepts came from the work of von Neumann, Eckert, Mauchly and others in the USA, as they had described them at that conference.

"Manchester built several big machines. This was not the case with Cambridge, which built EDSAC then EDSAC 2; but it was followed by TITAN – a Ferranti-built machine, based on the Manchester Atlas design. Their next mainframe was an IBM 370/195; so, it's fair to suggest that building large machines at Cambridge ended with EDSAC 2 – their last big computer. Roger Needham did lead a project to build a capability machine in the 1980s, while the work of Andy Hopper and others with the Cambridge Ring high-speed networking has also been influential.

"Eventually, the British mainframe computer industry, through a series of takeovers and mergers, ended up as one company – ICL. They had a large plant at West Gorton. In 1962, the Atlas – a most impressive machine – was Manchester's flagship. It was the world's first supercomputer, designed to execute a million instructions per second (MIPS). It had hardware floating-point and their patented virtual memory, which is everywhere now, even in mobile phones and personal computers. The Manchester carry chain came from there too. Its carry look-ahead adder is the logical way to perform calculations. Also, prior to Atlas, Manchester encoding was invented, which later became the basis of the IEEE 802.3 Ethernet signalling standard. The University of Manchester's last big machine was the MU5 – prototype for the ICL 2900 mainframe, which Inland Revenue relied on until the late 20th century. So, these big machines were highly influential.

"After I took up the ICL Chair of Computer Engineering, the university shifted its focus from designing big machines to chips through the AMULET research projects in the 1990s. Then with SpiNNaker, they returned to building big machines.

"Cambridge's contribution can be seen in the work of people like Andy Hopper at the university and the Olivetti Research Labs, where many brilliant developments have been extraordinarily successful."

00100011

Acorn Machines

From 1978, when Acorn Computers began developing a single board computer, until 1998 when it cancelled its most ambitious project the Phoebe, the company displayed a level of innovation seldom seen in a business of its size. This summary shows the evolution of its product development arranged into six categories:

Early Acorn Computers This began with the transition from consulting services to electronics design and manufacturing. While most of Acorn's products were on par with other manufacturers of the time, local area networking with the Econet was an exception. The end of this period is marked by the BBC evaluation of the Proton, which led to Acorn being chosen to produce the BBC Microcomputer in February 1981.

BBC Microcomputers and 8/16-bit spin offs This period began with the BBC Micro, followed by the lower-cost Electron, a version of the BBC Micro for the US and Canadian markets, various second processor options, the Master series and packages for office applications. One anomaly was the 16-bit Acorn Communicator.

32-bit computers Acorn re-set the company's direction when it developed its own microprocessor – the Acorn RISC Machine (ARM). This spawned new hardware for computer and communication applications, as well as critically significant operating system developments.

Acorn diversification This reflects Acorn's restructure to Acorn Network Computing, Online Media and Acorn RISC Technologies in 1995.

Acorn reunification To strengthen its image, Acorn kept its internal divisional structure but marketed all of them under the Acorn name.

Re-branded computers Acorn repackaged other brands (Olivetti and Psion) to meet specific marketing objectives. Neither is significant in the history of the company but are included here for completeness.

Early Acorn Computers

All products produced during this early period could be purchased fully assembled or as DIY kits.

Acorn Microcomputer Acorn's first off-the-shelf product was based on a single board computer, designed by Sophie Wilson while she was still at the University of Cambridge (the HAWK). The Acorn Microcomputer came with a 6502-processor, 256 bytes of static RAM and a 512-byte monitor program in PROM (later in masked ROM). *Released in 1979.*

System 1 With a hexadecimal keypad, seven-segment display and cassette interface attached to the Acorn Microcomputer, the System 1 was Acorn's answer to the Science of Cambridge MK14. *Released in 1979.*

Acorn 6809 This card was a plug replacement for the Acorn Microcomputer. It came with one kilobyte of RAM and two kilobytes of ROM (which handled the Acorn VDU, ASCII keyboard and printer). *Released in 1979.*

System 2, 3 and 4 The Acorn Microcomputer or an Acorn 6809 processor card could be fitted in a Eurocard backplane to build a System 2 or 3 (single height) or a System 4 (dual height) rack system, which also supports a combination of peripheral interfaces. *Released 1979–1980.*

System 5 The System 3 and 4 could also be shipped with a 2 MHz version of the 6502 and faster dynamic memory (DRAM). This is the same processor fitted to the BBC Microcomputer. *Released in 1980.*

Acorn Eurocards A range of Eurocard-format expansion cards were developed, including read/write and read only memory, disc interface, parallel and serial interface adaptors, laboratory interface, Econet, analogue input/output, video interface/ASCII keyboard, and a 6502 in-circuit emulator (ICE). They were usually mounted in a single- or dual-height Eurocard rackmount cabinet with a backplane and power supply. *Released 1979–1980.*

Acorn Atom Acorn's first home computer came with a 6502, 2-to-12 kilobytes of memory, integer-only BASIC (floating point optional) and operating system supplied in ROM. An Econet interface and a floppy disk drive could be added via an expansion port. *Announced in 1980.*

Acorn Proton Project name for a proof-of-concept prototype that was demonstrated to a team from the BBC. Designed and built using wire wrap at 4A Market Hill, Cambridge in the first week of February 1981, it convinced the selection panel that Acorn should build the

BBC Microcomputer. It demonstrated concepts that had been under development by Steve Furber, Sophie Wilson and Chris Turner during 1980 and 1981. *Never released.*

BBC Microcomputers and 8/16-bit Spin-offs

BBC Model A The cut-down version of the BBC Microcomputer came with a 2 MHz 6502 processor, 16k bytes of RAM and the machine operating system (MOS) in 16k bytes of ROM, which included the Cassette Filing System. *First shipped in 1982.*

BBC Model B The expanded version of the BBC Microcomputer contained all the Model A features, plus an extra 16k bytes of RAM (32k bytes total), a full set of connectors on the underside of the case (with associated interface chips), and sideways ROM expansions. It also supported three options: the Disc Filing System (DFS), Econet Network Filing System (NFS) and the speech system. *First shipped in 1982.*

Acorn Electron This was a low-cost cut down version of the BBC Model B, but with a series of expansion options that allowed it to be upgraded to almost the same specifications as the Beeb. It was slower, particularly in the higher resolution screen modes. Supply problems delayed its release until after Christmas 1983, resulting in heavy losses to Acorn. It contributed to Acorn being taken over by Olivetti. *Announced in 1983.*

BBC Model B+ When Hitachi changed the format of dynamic RAM chips, the Model B+ served as a stopgap while the BBC Master was being developed. There were two models: One was equipped with 64k of memory and DFS. It was the first of the Acorn machines to come with a double density disk system as standard. The other also came with an extra 64k of sideways RAM. Several Model B glitches were fixed, and it was powered by a CMOS 65C02 chip. *First shipped in 1984*, it was soon superseded by the BBC Master. Acorn was criticised for releasing the machines so near to the Master's release.

Model B (USA/Canada) Designed to meet FCC regulations, produced by Wongs in Hong Kong, this BBC Model B variant came in a metal-lined case with DFS, Econet and speech systems, and NTSC-compatible TV adaptor. The *UK command was required to run some software. *First shipped in 1984.*

Cheese Wedges There were nine BBC Micro upgrades housed in this case, which was the same shape, but about half the width of a BBC Micro.

They included the Teletext Adapter, Prestel Adapter, IEEE-488 Interface, Econet Bridge, 6502 Second Processor, Z80 Second Processor, NS32016 Second Processor (a.k.a. Cambridge Co-processor), Universal Second Processor unit (which allowed earlier BBC Micros to access the same second processor features as the BBC Master), and the ARM Evaluation Unit. Acorn also produced 6502 Turbo Second Processors, with RAM expanded from 64k to 256k bytes. *First shipped between 1984 and 1986.*

Reuters APM Board The Application Processor Module was a custom-built system developed by Acorn for Reuters in the early-to-mid 1980s. It ran BCPL applications in demanding environments like trading floors, where standard BBC Micros weren't capable. It was a BBC Model B+ in Eurocard's large format for rack mounting. *First shipped in 1984.*

Acorn Business Computer Acorn's first entry into business machines – the ABC range – was cancelled before any were shipped. It consisted of a BBC Model B+ with a second processor packaged in a new case with an integral monitor and separate keyboard. Second processors included a 32016 running PANOS, 80186 running DOS and a Z80 running CP/M. Eight models were *announced in 1984.* They included the ABC Personal Assistant, ABC Terminal, ABC100, ABC110, ABC200, ABC210, ABC300 and ABC310.

Cambridge Workstation The ABC 210 was the only one released. The Cambridge Workstation came with a 32016 second processor, PANOS or Xenix and a 10-megabyte hard drive. *First shipped mid-1986.*

British Telecom Merlin M2105 Designed as a smart terminal; it was an Electron with a speech synthesiser and modem. *Announced in 1985.*

Communicator Another communication machine, it was a sleek, wedge-shaped business terminal designed for online communications, Prestel access, and enterprise messaging. It's one of the few Acorn machines to feature a true 16-bit CPU, the GTE G65SC816P-2, a variant of the 6502-processor. It came with an on-board modem, communications software, dynamic RAM plus 32k of CMOS RAM. It usually came bundled with office productivity software, depending on its target vertical market or application. *Announced in 1985.*

BBC Master Series Successor to the BBC Model B, they included a lot of software and hardware improvements, and were powered by the WDC 65C02 chip, which had extra instructions to improve performance. The basic machine was the **BBC Master 128**, which featured the Advanced Disc Filing System (ADFS) and 128k of memory as standard. A smaller footprint version was called the **BBC Master Compact**. Externally, the **BBC Master Turbo** and **BBC Master 512** appeared identical to the BBC Master 128, but with an internal second processor.

The Master Turbo was fitted with a 65C02 second processor. The Master 512 had an 80186 processor with DOS loaded from disk. Other variants include the **BBC Master AIV** (interactive video machine used in the BBC Domesday Project) and the **Master ET** (Econet Terminal). The BBC Master Series was *released in 1986 and remained in production until 1993.*

Briefcase Communicator This contains the same hardware as the Communicator repackaged in a briefcase, but with an extra RAM interface card. Early shipments went to Pickfords Travel and some government departments. *Announced in 1987.*

32-bit Computers

Archimedes Family This was launched in June 1987 as Acorn began to move away from its long-standing ties to the 8-bit microprocessors. The series of machines that followed would include: A305, A310, A410, A410/1, A420, A420/1, A440, A440/1, A540, A3000, A3010, A3020, A500, A5000, A680, M4, R140, R225, R260, A7000, A7000+

A500 Inspiration for the A300/A400 series of machines, it was fitted with the ARM2 processor and the older VIDC1 chip. *Never released.*

A3xx The BBC-branded first range of Archimedes computers was powered by a custom designed chipset based around the ARM2 RISC chip. They came with the Arthur operating system and ADFS in 512k bytes of ROM, which was little more than BBC MOS for RISC. While more advanced than the 8-bit machines they were replacing, their performance fell far short of their host hardware – but it was a start. The A305 came with 512k bytes of RAM and the A310 was fitted with one megabyte. They can be easily identified by their red function keys. *Announced in 1987.*

A4xx Released at the same time as the Archimedes A3xx series, the Acorn-branded A4xx series featured an in-built hard disk controller. Some models were supplied with 43-megabyte hard disk drives, and the series was capable of handling four expansion cards including a co-processor card. The A410, A420 and A440 came with one, two and four megabytes of RAM, respectively. They can be identified by their grey function keys. *Announced in 1987.*

A680 The development machine for RISC iX was different from most

Archimedes machines in that it did not have RISC OS in ROM and could only use a high-resolution monochrome monitor for display. It came with eight megabytes of RAM and a 67-megabyte SCSI hard drive by default. *Never released.*

M4 Another development machine for RISC iX, this was an A680 in a large case about eight times the size of a standard A400 machine. Only one of this model is believed to exist. It had eight megabytes of RAM and used a SCSI connection to a 67-megabyte hard drive. *Never released.*

R140 This was Acorn's entry-level machine into the Unix market. Its main drawback was that the ARM2 was not ideal for running Unix. Functionally identical to the A440, it had four megabytes of RAM and a 52-megabyte hard drive, with Acorn RISC iX. Due to memory limitations and a large page size, it did not fare well as a Unix workstation. *Announced in 1988.*

A3000 This was the first machine to be released with the newer and faster MEMC1a chip as standard, and the newer RISC OS. Designed to be an entry-level machine, it had only one full expansion port with a single internal mini expansion card slot, providing limited expansion. This machine was popular with schools as a low-cost entry point into the Archimedes machines. *Announced in 1989.*

A4xx/1 This was functionally identical to the 400 series but featured a newer memory controller, the MEMC1a, which gave it a slightly more speed. They all came with HD mini-floppy drives. The A410/1 came with one megabyte of RAM. The A420/1 also came with a 20-megabyte drive and two megabytes of RAM. The A440/1 also came with 53-megabyte hard drive and four megabytes of RAM. *Announced in 1989*, shortly after the launch of the A3000.

A540 This machine came with four megabytes of memory, an ARM 3 processor, a 100 Meg SCSI drive and a later version of RISC OS that was updated to handle up to 16Mbytes of memory. *Announced in 1990.*

R260 Functionally identical to the A540, the R260 came with Acorn's Unix (RISC iX), which ran much better due to its increased processing power. *Announced in 1990.*

R225 A diskless version of the R260, it was designed to be a cheap networked Unix station. Released at the same time as the R260. *Announced in 1990.*

A5000 This was the first machine to feature quad density 1.6-megabyte disk drives as standard and RISC OS Version 3 and ADFS. The high-end models also came with a 40-megabyte IDE hard drive (later increased to 80 megabytes). While it was released almost a year before Acorn's first portable, its packaging was an expanded version of the A4, on whose design it was based. *Announced in 1991.*

A4 Functionally the same as the A5000, the A4 was Acorn's first portable machine. Advanced for its time, it featured power saving modes and an LCD screen. It was called the A4 because its footprint was the same size as an A4 sheet of paper. *Announced in 1992.*

A30x0 The A3010 and A3020 were designed as low-end home computers. They were among the first to feature the ARM250 processor, which offered better performance than the ARM2, but less than an ARM3. It supplemented the A3000. They came with one megabyte of memory. The A3010 had high density drives and was the first to feature joystick ports. The A3020 was fitted with a 60-megabyte hard drive and a network connector. Due to supply problems with the ARM250 chip, early models came with a daughterboard that contained an ARM2, IOC, MEMC and VIDC which gave the equivalent of an ARM250. This was discontinued once supplies became more reliable. *Announced in 1992.*

A4000 This was a more expandable version of the A30x0 series, pitched toward the home office and more serious applications. Its three-box format (like the A5000) was also driven by an ARM 250 processor, but came with two megabytes of memory, an 80-megabyte IDE drive, plus optional Ethernet and Econet ports. *Announced in 1992.*

A5000 (alpha variant) This was an improved version of the A5000 with a faster (33 MHz) processor and could also support a faster floating-point arithmetic (FPA) chip. It came with 2 megabytes of RAM and an 80-megabyte hard drive or 4 megabytes of RAM and a 160-megabyte drive. Both configurations could be expanded to 8 megabytes of memory. *Announced in 1993.*

Acorn Risc PC series This was the next generation of machines - superseding, but compatible with, the Archimedes range. They featured configurable and modular systems with a great variety of options that were founded on the second-generation chipset with VIDC20, IOMD, the newer ARM6, and better cell processors.

Risc PC 600 series This machine launched the Risc PC range with the ARM 610 processor, memory storage ranging from two to eight megabytes of RAM, a mix of ADFS high density drives and IDE hard disks. The new machines featured the processor card option – the concept of which was first shown in the A540 – and a unique second processor slot that allowed the machines to have two different types of processors in the system at the same time. By fitting a 486 board, Intel-based software could be run on the machine adjacent to native ARM programs. Both processors could be allocated memory and shared the system and other resources. Memory management was also improved. The podule interface was also extended with DMA, extended addressing, 32-bit data pathways from the I/O system, and an expanded memory map for each podule. Realtime video from the I/O system became feasible with high-speed data transfer. *Announced in 1994.*

A7000 Successor to the A4000 machine, this was a cut down Risc PC for school/home entry. This and the Risc PC 700 were the first machines to feature RISC OS 3.6. Unlike the Risc PC the A7000 has no second processor slot, only one DRAM socket and no VRAM capabilities, restricting memory expansion to 128 megabytes plus the memory soldered on the motherboard. The A7000 Net had an Ethernet interface in place of the standard 425-megabyte hard drive. *Announced in 1995.*

Risc PC 700 series The second generation of Risc PCs to be shipped, they featured the newer ARM 710 processor, 16-bit sound with audio mixer on the motherboard and RISC OS 3.6. It was an incremental improvement of the Risc PC to keep pace with market changes. *Announced in 1995.*

Acorn Diversification

Network Computer This was Acorn Network Computing's implementation of the Oracle Reference Standard for Network Computers (NC). The standard NC came with four megabytes of RAM, and its only storage device was a smart card, but could be upgraded by the customer. *Announced in 1996.*

Set Top Box 1 Online Media first produced these as base units for the Cambridge Cable and Online Media trial. They were sold commercially, with two megabytes of RAM, four megabytes of operating system in ROM, a hardware MPEG card, and an infrared remote control. These were the first to target the new Digital TV market. *Announced in 1996.*

Set Top Box 20 Online Media developed an advanced network computer that provided interactive home TV. They required a high-capacity network, and featured hardware MPEG decoder and infrared remote control. It blurred the line between computers and home appliances. The two variants – one with CDROM drive and the other without – could connect to ATM or Ethernet networks using a podule. *Announced in 1996.*

Stork Acorn Risc Technologies released this successor to the A4 – a portable Risc PC that included a PCMCIA interface. It was the first Acorn portable to offer a docking station. Not launched as an Acorn product, it was licensed to volume customers. As a result, the LCD screen and other features of the design were chosen by the customer. *Announced in 1996.*

NewsPAD Part of a European Union and Acorn Risc Technologies' broader Network Computer initiative, alongside devices like the Stork laptop and Set Top Box 22. Input was through the in-built touchscreen overlayed on the colour LCD display. It had two PCMCIA slots and 496 kilobytes of non-volatile RAM – compared to the usual 240 kilobytes. A docking station had an external monitor, parallel, serial, keyboard and mouse ports, along with a floppy disc drive. It was never commercially released but served as a technology demonstrator. *Announced in 1996.*

StrongARM Risc PC Acorn Risc Technologies developed this machine. It had similar specifications to the A700 series, with larger hard drives, RISC OS 3.7 and a DIGITAL StrongARM processor. *Announced in 1996.*

Acorn Reunification

Set Top Box 22 Almost identical to the Set Top Box 20, the CDROM was removed, an MPEG2 hardware decoding card was added, and rebadged Acorn, rather than Online Media. *Announced in 1997.*

A7000+ This was an updated version of the entry level A7000 machine. Keeping the same styling, it came with a hardware floating point processor and the ability to use EDO RAM. *Announced in 1997.*

StrongARM RiscPC This Alpha variant was the same as its earlier version but with a faster StrongARM processor. Another variant (essentially the same machine) was referred to as the J233 – the main difference being the software installed on the J233 included Java and a browser. *Announced in 1997.*

CoNCord This was an up-rated version of the NC with a DIGITAL StrongARM (SA110) processor, expanded memory and on-board Java. It came in a Tower configuration with a custom rounded case. Like other NC designs, the CoNCord features a ROM Card interface for upgrading the OS and infrared remote control. *Announced in 1997.*

DeskLite An attempt to enter the thin-client market, this was Acorn's last move from desktop computing to embedded and networked systems and was part of the same strategy that led to the Network Computer and Set-Top Box. It was to be licensed to other manufacturers, not mass-produced in-house. It targeted enterprises needing low-cost terminals for remote Windows access. DeskLite was based on the ARM7500FE with a stripped-down version of RISC OS, though it relied heavily on network booting and remote application delivery. *Announced in 1998.*

Phoebe 2100 Acorn's bold swan song was also known as the Risc PC 2. It was Acorn's ambitious successor to the Risc PC. Designed between 1996 and 1998, it aimed to modernize Acorn's desktop line with cutting-edge features and a striking yellow tower case. It featured a 233 MHz Intel StrongARM SA110 (with plans for 300–360 MHz variants), up to 512 megabytes of SDRAM, IOMD2 (advanced I/O controller with multi-CPU support), VIDC20R (enhanced video controller with double the performance, four PCI slots, three Acorn Podule slots, games port, MIDI, IrDA, and Soundblaster-compatible audio, EIDE with support for four devices, CD-ROM, floppy, in an NLX-style tower with custom yellow front panel and RISC OS 4.

Despite working prototypes and developer previews, Acorn cancelled the Phoebe project as the company was in the process of winding up, as it moved to realise the greater value of its ARM Holdings shares. Only two machines were fitted with completed IOMD2 chips. One resides at the Centre for Computing History, the other at the National Museum of Computing. These two examples stand as a reminder of what could have been: a machine that embodied the peak of Acorn's technical prowess. *Cancelled in 1998.*

Re-branded Computers

Acorn M19 While not a product of Acorn R&D, the M19 was a strategic move to offer MS-DOS compatibility. It was used in education, government, and business environments where PC compatibility was essential. This was a re-badged Olivetti M19 PC-clone. Based on an 8088 processor it also came with its own monitor. Acorn bundled additional software and Econet networking, distinguishing it from Olivetti's standard offering. *Announced in 1986.*

Pocket Book This was an example of Acorn's strategy to extend its educational reach beyond desktops. They were rebranded Psion Series 3 – a compact, clamshell-style personal digital assistant (PDA) designed for the education market. Inbuilt applications included a calendar/organiser and spelling checker/thesaurus. It could link to Acorn, Mac and Windows computers. Acorn positioned it as a student-friendly digital organiser. Its portability, long battery life, and compatibility with school systems made it a practical tool for learning environments. *Announced in 1992.*

The Pocket Book II was an improved model based on the Psion Series 3a, which came with a faster NEC V3 processor. *Announced in 1993.*

SchoolServer In the 1990s, internetworking different computer platforms in the education sector became widespread – typically Acorn, PC and Apple Macintosh. This led to the introduction of the SchoolServer in 1995. It consisted of an IBM server running Windows NT 3.5, with a PowerPC processor, 24 to 32 megabytes of RAM, one or two 1 GB hard drives, an Ethernet interface, and bundled third party software to connect Acorn computers to the SchoolServer facilities. It was based on Microsoft's own platform and proprietary networking technologies. Acorn even became a Microsoft Solution Provider despite having been a vocal critic of Microsoft in the past. There was also the Super Server project reportedly the product of research at Oxford and Cambridge universities. The solution was deployed at several sites. *Announced in 1995.*

Acknowledgements

In his chapter, Carl Dellar recalls the words of one of the giants of UK computer research and innovation, who told him:

"I've written a few books myself, and I've found the best way is not to start at the beginning or at the end. Start with what you know most about. Write about that first; and whatever you do, write every day, because even if you cross it out later, you'll have gone through that process. It becomes a habit and eventually, you'll get there."

Professor Sir Maurice Wilkes FRS FREng FBCS, University of Cambridge

I am grateful for the support of everyone who assisted in this quest. To those who tolerated my daily requests for information, no words can adequately express my appreciation: Doug Pauling, Steve Furber, Chris Turner, Carl Dellar, Brian Cockburn, David Allen and Richard Russell – your advice and patience helped me bring this story to life.

I'd particularly like to mention: Andrew Gordon, Sir Andy Hopper, Arthur Norman, Christopher Curry, Colin Priestley, David Bell, Graham Tebby, Hugo Tyson, Jeremy Wills, Jim Mitchell, Joe Dunn, John Cox, John Radcliffe, Jon Thackray, Kim Spence-Jones, Laurence Hardwick, Lesley Curry, Mel Pullen, Paul Bond, Paul Fellows, Ramanuj Banerjee, Sam Wauchope, Sheila Innes and Sophie Wilson. The loyalty and support shown to one another during the Acorn years is still there today.

Thousands of people worked with Acorn Computers and the BBC. Some of the people I remember fondly have passed. There are some I'd have liked to interview but couldn't reach. To those not named, your contributions to Acorn's success were just as valuable. I accept sole responsibility for errors. I blame the tyranny of time, distance and fading memory for omissions.

Acorn and the BBC achieved what they did because people really cared and understood that what they were doing was important.

Rob Napier, 6 January 2026